China's Water Warriors

China's Water Warriors

Citizen Action and Policy Change

With a New Preface

ANDREW C. MERTHA

Cornell University Press   *Ithaca & London*

First published 2008 by Cornell University Press
First printing, Cornell Paperbacks, 2010

Printed in the United States of America

Library of Congress Cataloging-in-Publication Data

Mertha, Andrew, 1965–
    China's water warriors : citizen action and policy change / Andrew C. Mertha.
        p. cm.
    Includes bibliographical references and index.
    ISBN 978-0-8014-4636-8 (cloth : alk. paper)
    ISBN 978-0-8014-7668-6 (pbk. : alk. paper)
    1. Water resources development—Political aspects—China   2. Environmentalism—
Political aspects—China.   3. Economic development—Environmental aspects—China.
4. Environmental policy—China.   5. China—Politics and government—1976–2002.
6. China—Politics and government—2002–   I. Title.
    HD1698.C5M47 2008 2010
        333.9100951—dc22

2007036946

Cloth printing   10 9 8 7 6 5 4 3 2 1
Paperback printing   10 9 8 7 6 5 4 3 2 1

To my parents

Sok Szeretettel

Contents

Tables and Figures

Tables

Figures

Selected Institutions and Abbreviations

Archaeology Research Office	ARO	*Kaogu yanjiu suo*	考古研究所
Archaeology Brigade	AB	*Kaogu dui*	考古队
Black Dragon River Water Commission	BDRWC	*Heilongjiang shuili weiyuanhui*	黑龙江水利委员会
China Youth Daily	CYD	*Zhongguo qingnian bao*	中国青年报
Cultural Relics Management Bureau	CRMB	*Guojia wenwu guanli ju*	国家文物管理局
Cultural Relics Bureau (local)	CRB	*Wenwu ju, wenwu ke*	文物局，文物科
Cultural Relics Department (local)	CRD	*Wenwu chu*	文物处
Culture Bureau (local)	CB	*Wenhua ting*	文化厅
Culture Station	CS	*Wenhua zhan*	文化站
Datang		*Zhongguo datang jituan gongsi*	中国大唐集团公司
Development and Reform Commissions (local)	DRC	*Fazhan gaige weiyuanhui, fagaiwei*	发展改革委员会，发改委
Energy Bureau		*Nengyuan ju*	能源 局
Environmental Protection Bureaus	EPB	*Huanbao ju*	环保局
Friends of Nature	FON	*Ziran zhi you*	自然之友
Green Watershed	GW	*Yunnan sheng dazhong liuyu guanli yu yanjiu zhongxin*	云南省大众流域管理与研究中心
Guodian (Group) Corporation	GDGC	*Zhongguo guodian jituan gongsii*	中国国电集团公司

Hongta Group	HTG	*Yunnan hongta jituan*	云南红塔集团
Huadian (Group) Corporation	HDGC	*Zhongguo huadian jituan gongsi*	中国华电集团公司
Huai River Water Commission	HRWC	*Huaihe shuili weiyuanhui*	淮河水利委员会
Huaneng (Group) Corporation	HNGC	*Zhongguo huaneng jituan gongsi*	中国华能集团公司
Huaneng Lancang River Corporation	HLRC	*Huaneng Lancang jiang gongsi*	华能澜沧江公司
Irrigation and Water Conservancy Department		*Nongtian shuili chu*	农田水利处
Jianshe (Construction) Bureaucracy		*Jianshe xitong*	建设系统
Jiaotong (Communication) Bureaucracy		*Jiaotong xitong*	交通系统
Ministry of Culture	MOC	*Wenhua bu*	文化部
Ministry of Water Resources	MWR	*Shuili bu*	水利部
Ministry of Water Resources and Electric Power	MWREP	*Shuili dianli bu*	水利电力部
National Development and Reform Commission	NDRC	*Guojia fazhan gaige weiyuanhui, fagaiwei*	国家发展改革委员会，发改委
Nu River Project	NRP	*Nujiang dashuiba gongcheng*	怒江大水坝工程
Pearl River Water Commission	PRWC	*Zhujiang shuili weiyuanhui*	珠江水利委员会
Sanxia (Three Gorges)		*Sanxia jituan gongsi*	三峡集团公司
Songliao Water Commission	SLWC	*Songliao shuili weiyuanhui*	松辽水利委员会
State Environmental Protection Administration	SEPA	*Guojia huanbaoju*	国家环保局
State Investment and Production Management Commission	SIPMC	*Guojia zichan guanli weiyuanhui*	国家资产管理委员会
State Power Company of China	SPCC	*Guojia dianli gongsi*	国家电力公司
Taihu River Basin Management Bureau	THRBMB	*Taihu liuyu guanli ju*	太湖流域管理局
The Nature Conservancy	TNC	*Meiguo daziran baohu xiehui*	美国大自然保护协会
World Heritage Office	WH	*Shijie yichan bangongshi*	世界遗产办公室
Yangtze River Water Commission	CJWC	*Changjiang shuili weiyuanhui*	长江水利委员会
Yellow River Water Commission	YRWC	*Huanghe shuili weiyuanhui*	黄河水利委员会
Yunnan Hydropower Group	YHCG	*Yunnan shuidian jituan*	云南水电集团

Preface to the Paperback Edition

In 2008, when *China's Water Warriors* first appeared, the events depicted were very recent, spanning 2003–2006. Two years since publication may not seem very long, but in China, with its rapid shifts in political mood and breathtaking economic growth, a lot can happen in twenty-four months. How do the book's conclusions stand the test of time?

The good news is that the three empirical case studies continue along the same trajectory I originally reported. Moreover, political pluralization—the theoretical framework suggested by my research—shows promise as a way to understand various other policy areas in China, not simply hydropower or the environment. The bad news is that the political climate in China has chilled considerably since I undertook my initial research. This change has constrained the ways in which non-traditional actors have been able to enter and influence the policy process, particularly at the local level. As a result, there have been fewer events than I anticipated to test the hypotheses generated by *China's Water Warriors,* and I have tempered my optimism for political liberalization in the short-to-medium term.

As far as the three cases—Dujiangyan, Pubugou, and the Nu River Project—are concerned, there have been few new developments since 2008. In the case of Pubugou, Hanyuan county, Sichuan province, the project has moved forward while widespread opposition to it has been neutralized by the coercive arm of the state, even though few if any of the issues raised by the protesters have been addressed, let alone resolved. As I describe in chapter 4, in the fall of 2004 widespread opposition by people in Hanyuan county led to some of the largest demonstrations in the history of the People's Republic of China. Hanyuan residents objected to leaving their homes, their land, and their livelihoods and, for shockingly low compensation, re-settling in areas with expensive but shoddily built housing, poor land, and neighbors who viewed them as unwelcome troublemakers. Since 2006, only a very small group of these resettled people (*yimin*) have seen their lives improve thanks to opportunities not available to everybody, such as out-side business ventures or securing work as migrant laborers further afield. For most *yimin*, the socioeconomic situation has not improved. While some have managed to maintain their previous standard of living, far more have seen a drop in their living standard: as they had feared, the allotted land and housing in the resettlement areas are poor and cannot be used to pro-duce crops of the same quality.

There are other problems. Although 30,000 or so *yimin* were resettled in mid-2006 to counties outside of Hanyuan, others were relocated within the county. These people have been moved to sloping plots of land (around the new county seat of Luoshigang) which are not only poorly suited for grow-ing crops but also require special building materials and design expertise well beyond what the new residents can afford. To add injury to insult, in August 2009, as a new road above the water level was being built to replace one that was to be submerged, a large section of the road on an unstable section of the mountainside broke off, killing motorists and people sleep-ing in their homes below. The official number of those killed was "more than ten" (*shijge ren*) people, although the real number is more like several dozens of people (*jishige ren*).

Yet some people still refused to move. In 2009, the water level of the res-ervoir reached 790 meters, and in 2010, it reached its highest level, 850 me-ters. In the intervening year, more than one hundred people who refused to move voluntarily were forcibly relocated by the Public Security Bureau (*gong'an*), the People's Armed Police (*wujing*), and the Fire Brigade (*xiao-fang dui*). Resistance, however, was isolated and did not evolve into any sort of organized opposition. The story depicted in *China's Water Warriors* re-mains what it was: apart from delaying the project for a year, the 2004 dem-onstrations in Hanyuan did not prevent the Pubugou Dam Project from going forward as originally intended.

The case of Dujiangyan remains, like Pubugou, also largely unchanged. There have been no attempts to revive the Yangliuhu Dam Project, which opponents argued would have dire negative consequences for the 2,250-year-old Dujiangyan Irrigation Works. Indeed, given the proximity of Du-jiangyan to the epicenter of the May 12, 2008, Wenchuan earthquake, which measured 8.0 on the Richter scale, the focus has been not on undertaking new projects but on repairing old ones. The most significant of these is the Zipingpu Dam, fewer than ten kilometers upriver from Dujiangyan. In *China's Water Warriors,* Zipingpu plays a supporting role as the dam project that, though opponents refrained from protesting against it in 2001, emboldened them in 2003 to mobilize against Yangliuhu.

But today Zipingpu represents a larger potential threat to Dujiangyan than Yangliuhu ever could. One of the effects of the Wenchuan earthquake was to reshape the land under Zipingpu's foundation. In addition, the earthquake caused major fissures in the dam itself, cracks that are clearly visible to the naked eye. Although the authorities have since reduced the water level, the 156-meter-high dam still contains an enormous amount of water exerting tremendous pressure. The shift in the geology of the Longmenshan area has also shifted the direction of the water pressure. If the dam is breached, 1.126 billion cubic meters of water would destroy everything in its path. The water would obliterate Dujiangyan, ruin Sichuan province's crop yield, and kill hundreds of thousands, even millions, of people. Some scholars have even suggested that Zipingpu was instrumental in *causing* the Wenchuan earthquake.[1] Although speculative, that claim echoes warnings issued during the planning of this and other dam projects. Most troublesome is the fact that these potential dangers have not slowed down the pace of construction elsewhere in China.

The earthquake prompted perhaps one consolation. It forced the government to provide the thousands of *yimin* who had been moved out of their homes in order to construct Zipingpu, but who had not yet been provided with new housing, the same levels of compensation as those given to the earthquake victims in the countryside. That aside, the case of Dujiangyan documented in *China's Water Warriors* remains the same.

The third case, the Nu River Project (NRP), appears to be as far away from resolution today as it was in 2004. As was the case when I was researching and visiting the upper reaches of the Nu in 2005 and 2006, numerous rumors are swirling. Many of them revolve around the idea that although the project has been put on hold, preparations are still being made in anticipation of a go-ahead from Beijing. However, accounts vary about which of the thirteen sites for hydropower stations are being prepared. Something is certainly taking place in Liuku municipality, at the halfway point of the

1. See Sharon LaFaniere, "Possible Link between Dam and China Quake," *New York Times,* February 5, 2009; and Malcolm Moore, "Chinese Earthquake May Have Been Man-Made, Say Scientists," *Telegraph,* February 2, 2009.

Nu as it snakes along a north-south axis, but there remains considerable debate about other spots along the river. Back in 2006 there were preparations as far north as just inside the Yunnan-Tibet border in the Songta section of the project. Some have said that preparations are also under way in Lushui, Maji, and possibly elsewhere. But, rumors aside, the deadlock that emerged from the mobilization both for and against the NRP remains almost eight years on.

As inductive case studies, Pubugou, Dujiangyan, and the Nu River Project can only suggest theory and attendant hypotheses; they do not—indeed, methodologically they cannot—test them. But events such as the 2005 environmental protests in Dongyang municipality, Zhejiang province, as well as the events surrounding the Xiaonanhai Dam Project on the Upper Yangtze just upriver from Chongqing municipality, among few others, do provide opportunities to test the claims raised by *China's Water Warriors*.[2]

If the empirical studies more or less retain their integrity, what about the theoretical framework that they suggest, that of political pluralization in the policy process with the inclusion of non-traditional actors (disgruntled local or peripheral officials, the media, and non-governmental organizations)? This area of inquiry shows particularly encouraging developments because even within the generally inert political status quo in China, a growing number of policy areas exhibit the same political dynamics at work, with very similar processes and outcomes.

My argument that policy entrepreneurs manipulate issue frames in order to mobilize support, constrain government pushback, and link with other interested actors applies to areas outside of hydropower as well.[3] I have recently shown how these same dynamics were at play in the case of the Rifeng Lighter Factory in Wenzhou municipality. The company chairman, Huang Fajing, used issue frames drawing from the language of the Korean War (or the "Resist America, Aid Korea" War) to mobilize support initially within the media and ultimately among Chinese government officials and interested parties against what he saw as unfair trade practices that targeted his industry. The result was China's first initiation of a trade dispute as a member of

2. On Xiaonanhai, see, *inter alia*, "New Yangtze Dam May Be Death Sentence for Rare Fish," *Reuters*, June 19, 2009.

3. John Flower makes an excellent point when he asserts that historical issues were also at play in the oppositional narrative at Pubugou. (I had argued that opposition rested on bread-and-butter compensation claims.) That said, issue frames are only a necessary condition, not a sufficient one, and they were quickly overwhelmed by events in the form of demonstrations and large-scale protests, which essentially made these alternative narratives moot. See John Flower, "Ecological Engineering on the Sichuan Frontier: Socialism as Development Policy, Local Practice, and Contested Ideology," *Social Anthropology* 17 (1) 2009: 40–55.

the World Trade Organization.[4] Huang's strategy has since been adopted by other entrepreneurs in Wenzhou and elsewhere. Other analysts have suggested similar dynamics in other policy areas. Yawei Liu puts it this way:

> If "warrior" is meant in a more abstract sense—as a label given to those Chinese citizens who have managed to force the government at various levels to reverse a decision or who have created such social momentum that a certain claim must be disputed—then there are too many such warriors in China to count. "Election warriors" in Shenzhen and Beijing ran as independent candidates in 2003 and won. "Zoo warriors" and "imperial garden warriors" in Beijing managed to reverse the decision to relocate the zoo and to develop the ruin of Yuanmingyuan. "Internet warriors" in Yunnan used the Internet to overturn a provincial government's claim that an inmate died by accident when playing hide and seek.[5]

Although this plays just a bit fast and loose with the framework that I introduce in *China's Water Warriors,* Liu is devastatingly accurate when he says that the government brooks little opposition in framing media coverage to curb social instability,[6] as was amply demonstrated in the coverage of the Wenchuan earthquake. Indeed, even though stories abounded about poorly built schools crushing middle school students (builders had substituted wood for steel reinforcements, I was told by survivors) while government buildings survived intact, these stories were quickly brought under control. I was shocked when some of the people in the affected area in Pengzhou county told me that government officials fled as soon as the earthquake subsided and people went days without any kind of governing structure until People's Liberation Army (PLA) units were able to reach Pengzhou.[7] Had the government allowed such stories to get out at the time, public opinion inside and outside China might have been quite different.

4. Andrew Mertha, "Fragmented Authoritarianism 2.0: Political Pluralization of the Chinese Policy Process," *China Quarterly* 200 (December 2009): 1–18; and "Society in the State: China's Nondemocratic Political Pluralization," in Peter Hays Gries and Stanley Rosen, eds., *Chinese Politics: State, Society and the Market* (London: Routledge, 2010), 69–84.

5. Yawei Liu, "China's Warriors Are Handicapped," National Bureau of Asian Research, *Asia Policy* 8 (July 2009): 140.

6. But even this point needs to be qualified. Opposition to the Xiaonanhai project, for example, was reported in a remarkably straightforward manner by the government-run English-language mouthpiece, the *China Daily.* (Certain official English-language media sources in China have become quite open in the reporting of stories that may embarrass the authorities and have even criticized the government.) See Wang Qian, "Experts Fight to Plug Up Dam Project," *China Daily,* November 12, 2009, http://www.chinadaily.com.cn/china/2009-11/12/content_8954067.htm.

7. Interviews with displaced residents in Pengzhou county, Sichuan, June 25, 2010. Accounts of how long it took ranged from one to three days, although this discrepancy may be

Liu's comments appear in roundtable proceedings published by the National Bureau of Asian Research (NBR), in which several scholars proffer observations and conclusions that go beyond the modest ones I make in *China's Water Warriors*.[8] David Lampton zeros in on the book's conclusion, specifically my claim that the pluralization of the policy process in China may produce constraints on decision making, thus leading to outcomes contrary to the goals of policy entrepreneurs and/or their supporters and allies. In short, pluralization could lead to negative unintended consequences. It is worth quoting Lampton at some length:

> If environmentalists and the pluralized policy process in China slow down the ability of Beijing to implement energy projects that do not depend on coal (hydro and nuclear projects, for example), this will force China to fall back on its default energy—coal. Coal increasingly is being recognized as having serious global (as well as domestic) externalities; the degree to which domestic political paralysis prevents China from moving away from coal will be the degree to which China has added opportunity for conflict with the international community.[9]

But Lampton does not stop here. He extrapolates to suggest an even broader point that I did not make explicitly but should have:

> As China becomes more pluralized and constrained by domestic fragmentation, the outside world may find it progressively more difficult to get PRC cooperation in a number of areas that inflict costs on well-organized constituencies within China. In short, not all good things necessarily come from pluralization....For Americans [the lesson] may be that China gradually is heading in the direction of a more responsive policy, a more constrained leadership; one cannot, however, expect that more pluralization will always translate into more international cooperation or Chinese behavior more aligned with US interests.[10]

Or, as Peter Ford succinctly puts it: "'Pluralization' clearly does not mean freedom," at least not for policymakers.[11] These analysts caution not only that pluralization and democratization are conceptually independent but that pluralization could represent a net loss of freedom for decision makers

---

due to the fact that some interviewed were residents of Pengzhou county, while others were residents of Pengzhou municipality.

8. National Bureau of Asian Research, *Asia Policy* 8 (July 2009), http://www.nbr.org/publications/element.aspx?id=389.

9. David M. Lampton, "Water Politics and Political Change in China," National Bureau of Asian Research, *Asia Policy* 8 (July 2009): 124–125.

10. Ibid., 125.

11. Peter Ford, "Breaching the Dam," National Bureau of Asian Research, *Asia Policy* 8 (July 2009): 144.

in China and could lead to perverse outcomes and consequences which are the exact opposite of those hoped for from these newly-mobilized actions.

Regardless of how we conceptualize political liberalization in China, there has been frustratingly little of it since 2007. To some degree this was to be expected, given Beijing's hosting of the 2008 Olympics and Chinese leaders' obsession with making the games go without a hitch. During an exchange on National Public Radio in early 2007, James Mann argued that China would liberalize until after the closing ceremonies and then clamp down again.[12] David Shambaugh, appearing on the same program, argued that this prediction got it exactly backward and provided a far more convincing argument that we were likely to see a restricted political space running up to the Olympics and then a return to the liberalizing trajectory. That is what should have happened, but it didn't. The political climate did not become less restrictive after the completion of the Beijing Olympics.

There were at least two events that neither Shambaugh nor I (nor, for that matter, China's leadership) anticipated. First was the Tibetan uprising, beginning on March 10, 2008. The second was the Uighur riots in Urumqi and Kashgar in July 2009. The Tibetan uprising had two effects that are relevant here. First, it led to an official state frame that many Han Chinese could use to make sense of the events that unfolded in Tibet, Western Sichuan, Qinghai, and elsewhere. Second, it had a dampening effect on prospects for political liberalization after the Olympics. Some of the coercive practices employed by the government were straight out of the Cultural Revolution playbook: struggle sessions, torture, thought reform, and the like. Recently a Tibetan monk released from custody following extensive thought reform provided a chilling account of the continuation of some of the darkest practices in Chinese politics.[13]

Anecdotally, I also experienced this change in political mood. In March 2006, I was able to travel into Songta, Tibet, largely unnoticed and completely unmolested, to observe the preparations under way for the Nu River Project. By contrast, in March 2009, I was planning on going to some of the areas devastated by the Wenchuan earthquake. Displaying a healthy dose of inattention, I happened to do so on the one-year anniversary of the Tibetan uprising. As a result, I was detained on the road for ninety minutes by armed and uniformed military and paramilitary officials while they tried

12. "China's Rise: A look at China's growing economic might and global influence, and US-China policy," *The Diane Rehm Show,* NPR, March 5, 2007, http://thedianerehmshow.org/shows/2007-03-05/chinas-rise.

13. Edward Wong, "After Chinese Re-Education, Monk Regrets Action," *New York Times,* June 29, 2010.

to make sure that I was not, in fact, a journalist. Despite the frustration on all sides in trying to prove a negative, these officials were perfectly cordial and professional. But it was also clear that they were under strict orders. When they were suitably convinced that I was who I said I was (or more accurately, when they felt that they could convince their superiors), I was allowed to return the way I had come.

The Uighur uprising a few months later reinforced the tendencies on the part of the Chinese government that have led to a newly restrictive political climate. By a strange coincidence, I happened to be in Urumqi and Kashgar in the days leading up to the first day of rioting. (I flew out of Urumqi at about the same time as the rioting got under way on the afternoon of July 5, 2009.) I was accompanying a congressional staff delegation, and we were meeting with local officials (including many within the coercive arm of the state, since we were discussing terrorism and drug trafficking, among other issues). What struck me was how unprepared these officials appeared to be for some sort of "event"—given the news of Uighur-Han clashes at factories in Guangdong the previous week, we sensed that all was not "pan-ethnic love," as they assured us. I was also struck with just how rigid and orthodox the discourse was. I did not expect Xinjiang to be a hotbed of political liberalism, but the rigidity and tone-deafness of our hosts was astonishing.

My more recent trips to China in early and mid-2010 have not suggested to me a softening of the political mood. This is perhaps not surprising either. In addition to the recent Tibetan and Uighur uprisings, we are in the early stages of the next succession struggle, which will culminate in the 18th Party Congress in 2012. To err on the side of creativity and to stray from political orthodoxy right now is widely understood as committing career suicide. If the past is any guide, it will probably take another five years before the new leadership is chosen and has consolidated control of the government, the Party, and the army. Only then are we likely to see a broad resumption of the trend toward greater political liberalization that is documented in *China's Water Warriors*. With a bit of luck, developments in China will undermine my prediction, as they have so many times before. From 2008 to 2010, however, although a small window to a bit more political pluralization may have opened following the Olympics, it was slammed shut by the crackdowns in Tibet and Xinjiang.[14]

China's Water Warriors may have been ahead of its time not because of any prescient gifts on my part but rather because of the cessation of liberalizing trends in China. This realization is disheartening in the short term, but the

---

14. Pluralization may have also fallen victim to China's need to put on its best face for the 2010 Shanghai Expo.

longer-term prospects are not terribly different from what I concluded in the first edition. I have no sense of a backsliding or reversal in pluralization; it is just an extended pause. As with so many developments in China, we can see this as "two steps forward, one step to the side." I feel rather pessimistic in the short term, but I remain optimistic for the long run.

Hong Kong, August 2010

Preface

In March of 2004, my colleague Bill Lowry, clutching that morning's *New York Times*, came into my office. A front-page article by Jim Yardley, "Dam Building Threatens China's 'Grand Canyon,'" had piqued his interest. I had already seen the article, and indeed, my first reaction had been a desire to run upstairs and show it to Bill. The article documented the grassroots attempts to reverse the decision to build a gargantuan hydropower project along the Nu River in remote northwest Yunnan province. Sure enough, a month later, in an article titled "China's Premier Orders Halt to a Dam Project Threatening a Lost Eden," the *Times* reported on Premier Wen Jiabao's decision to put the project on hold. As the conventional wisdom holds that Beijing enthusiastically pursues massive hydropower projects, with ministries and local governments falling into line, we regarded this as a noteworthy development. Together, we pooled our research funds and embarked on a trip to China that August to try to understand what was unfolding. In June, as I was conducting the preliminary research on the Nu River Project, I came across a similar event, indeed an even more dramatic case, in Dujiangyan, Sichuan province. By a stroke of good fortune, I had established contacts in Dujiangyan the previous summer—evidently I had performed satisfactorily during our "liquid lunch" as we were warmly welcomed back to Dujiangyan. We made that beautiful city the focal point of our research. The fruit of our labor was "Unbuilt Dams," an article that appeared in *Comparative Politics;* it forms the core of chapter 4 in this book.

I had originally conceived of this project as an interesting diversion from my research on bureaucratic (re)centralization in China, but I quickly found myself drawn to the fascinating world of hydropower politics. As a result, hydropower became the focal point of my research for the two and a half years following that initial visit. This book is the result.

My former advisor Ken Lieberthal and I have discussed many times how the politics of water reveal a great deal about how the Chinese state operates. This notion is borne out by the early literature on the creation and implementation of policy in China. Another of my advisers, Mike Oksenberg, wrote his doctoral thesis on irrigation during the Great Leap Forward. David Lampton's work on water policy provided the foundation to the groundbreaking 1987 volume he edited, *Policy Implementation in Post-Mao China*. And dams and hydropower provide one of the three cases in Lieberthal and Oksenberg's own seminal 1988 work, *Policy Making in China*. In the two decades since, the treatment of hydropower policy has been absorbed into the growing literature on environmental politics in China. *China's Water Warriors* returns us to water policy as a way of understanding how the policy process works in China more generally—particularly how it has changed since Lieberthal and Oksenberg conducted their research in the mid-1980s.

Over the past two decades, the "fragmented authoritarianism" (FA) framework identified by Lieberthal, Oksenberg, and others in the late 1980s and early 1990s has itself been applied to other policy areas, including the military, education, commerce, and trade. The FA framework remains remarkably durable.

However, new dynamics have emerged in the past few years that require us to recast our notions of how the political process works in China. Chinese scholars, activists, disgruntled local officials, media outlets, and even the broad public have become far more important to the policy process. Indeed, one of the most captivating dimensions of this research was this heterogeneity of political actors. Each of these actors, no matter where he or she stood on the actual policy question of whether project X should be built, was genuinely committed to that stance, and often it was difficult to extricate policy goals from deeper, more personal ones. Hydropower engineers argued convincingly that they were helping the country in its constant quest for energy sources. Local political leaders in favor of the projects understood them as ways to bring their own constituents out of desperate poverty. National-level officials, regardless of their views on particular hydropower projects, felt an obligation to allow interior provinces to share in the dramatic wealth that has made coastal China one of the most dynamic places on earth.

But it is the opposition—activists, officials, journalists, and scholars—that stands out. I did not always agree with their goals or their tactics—although my opinion is immaterial—for I was there as a social scientist, committed to maintaining a degree of objectivity that many of my sources found strange, even exasperating! However, I was and remain truly humbled by the purity of their commitment. These actors are venturing into new and largely uncharted political territory. Although as policy entrepreneurs—advocates for policy proposals or for the prominence of an idea in the policy realm—they have learned how to play the political game adroitly and effectively, the boundaries of acceptable political behavior are changing, precisely as a result of their actions. Their putative goal may well be to change not politics but policy. Distinguishing the two, however, is often controlled by precisely those parties against whom these activists are exerting their considerable energies. Their behavior is the very definition of political courage.

Sometimes the opposition fails, sometimes it succeeds, and sometimes the definition of success or failure remains elusive. In this book, I look at three cases. Chapter 3 traces the failure of the opposition to the Pubugou Dam in Sichuan province. Chapter 4 documents the dramatic success of the anti-hydropower forces at Dujiangyan, also in Sichuan. Chapter 5 analyzes the protracted power struggle between the two camps over the Nu River Project in neighboring Yunnan province.

What do these three stories tell us about hydropower politics? What do they tell us about China's political processes more generally? Most important, they demonstrate that these things *can* happen. Inspired by Robert Axelrod's *The Evolution of Cooperation,* his path-breaking work on international cooperation under the conditions of anarchy, I am documenting existence claims. Axelrod claimed that international cooperation can take place even under conditions of anarchy; I am making the claim that political pluralism can take place within—indeed, influence—the policy process in single-party, authoritarian China. In addition, even though my three cases cover roughly the same time frame and geographical area, there is considerable variation in the outcomes. This suggests certain larger hypotheses. Given the limitations of the case study method, I cannot claim that I am testing hypotheses here, but in terms of internal validity, these cases are quite robust.

Considering China's extraordinary complexity, I have found it helpful to maintain a requisite degree of humility when taking on the Middle Kingdom as the object of my scholarship. With that in mind, I hope I have been able to uncover some of the more interesting dynamics in that fascinating country's recent political trajectory.

In one end-of-term evaluation a few years back, an anonymous student, commenting on the riskier aspects of my field research, wrote that I "could do for political science what Indiana Jones did for archeology." While I certainly felt the temptation to don (if only metaphorically) the famous fedora many times in the field, Jones was basically a loner. I, on the other hand, have depended on the generosity of family, colleagues, friends, and strangers in order to accomplish what I have done, both in the field and outside of it.

Several people were instrumental in piquing my interest in hydropower politics. My colleague and friend Bill Lowry stands at the top of that list. Others include Ken Lieberthal, Liz Economy, Jim Yardley, and a host of individuals in China. In fact, this last group, a critical part of the story I tell, was extraordinarily generous in allowing me to enter its dense network of activists, journalists, and sympathetic officials. Although many of its members would not necessarily agree with some of the points made herein, this group quite simply made this book possible. That they must remain anonymous underscores the depth of my gratitude to interlocutors who, despite the risks inherent in doing so, helped me tremendously over the past thirty months.

Others in China provided the contacts necessary for me to meet with officials and scholars who are important players in the hydropower policy process. These include bureaucrats, engineers, scholars, and others who, while remaining unnamed here, were crucial to allowing this story to be as accurate and balanced as possible.

Scholars in the United States were extremely generous with their time. They read sections of the manuscript or listened to presentations of parts of the argument in professional conferences, in seminars, even in hallways, offering helpful suggestions and comments. Starting with my colleagues at Washington University, I extend my thanks to Randy Calvert, Bob Hegel, Nate Jensen, Mona Krook, Bill Lowry (again), Steve Miles, Gary Miller, Itai Sened, Andy Sobel, and Jim Wertsch. Many in the Chinese politics community have been equally generous with their comments and critiques. My thanks go to Joe Fewsmith, Mark Frazier, Mary Gallagher, Scott Kennedy, Pierre Landry, Ralph Litzinger, Ken Lieberthal (again), Kevin O'Brien, Dorie Solinger, and Susan Whiting. Outside the China community, I thank James Kuklinski and Olga Shvetsova for their excellent suggestions.

In the United States and China, many individuals made it possible for me to undertake field research at a degree of comfort and predictability that I had no right to expect. For this I am very grateful. I thank He Lihua, Lan Lan, Mindy Liu, Xiao Zhu, Yang Hong, and Yang Wei. Graduate students who provided yeoman service include Liu Hua, Dan O'Neill, Pang Xun, and Xu Xin. Other individuals at varying stages of their careers have also been extremely helpful. I thank Darrin Magee, Beth Kinne, Kristen Mac-Donald, and Travis Winn for sharing their insights with me. Fellow intrepid traveler John Kandel held down the fort in Chengdu during my trip to Hanyuan in August 2006.

Steve Smith and the Weidenbaum Center on the Economy, Government, and Public Policy at Washington University have made this research possible through a large series of small grants.

Roger Haydon has demonstrated once again his considerable talents as an editor. He and two anonymous reviewers have allowed the germ of an idea to evolve into a book that is a distinct improvement over the initial manuscript by several orders of magnitude. I am indebted to all of them, as I am to Karen Laun and the whole team at Cornell University Press.

The extraordinary degree of assistance provided by all these individuals notwithstanding, some errors inevitably remain, for which I am solely responsible.

Finally, my family has been tremendously supportive. My wife Isabelle and my daughter Sophie have allowed me the time necessary to complete this project. They demonstrate to me every single day that there is life outside of scholarship—and what a life it is!

Unbeknownst to them, my parents provided me at an early stage with the skills necessary to complete this project. My father, Gustav Mertha, instilled a love of travel and an acceptance of hardship on the road that allowed me to brush off inconveniences, like delays caused by mudslides on the roads in northwest Yunnan, with a shrug. My mother, Eva Foreman, showed me the rewards of putting one's nose to the grindstone and the importance of pure, hard work in achieving positive results. Their escape from Hungary during the 1956 revolution and their audacity in establishing a new life in the United States inspired me to take the road less traveled, a road that ultimately brought me to China. It is to them that this book is dedicated.

China's Water Warriors

1 | China's Hydraulic Society?

> Man never stops affecting his natural environment. He constantly *transforms*
> it; and he *actualizes* new forces whenever his efforts carry him to a new level
> of operation. Whether a new level can be attained at all, or once attained,
> where it will lead, depends first on the institutional order and second on the
> ultimate target of man's activity: the physical, chemical, and biological world
> accessible to him. Institutional conditions being equal, it is the difference
> in the natural setting that suggests and permits—or precludes—the
> development of new forms of technology, subsistence, and social control.
>
> —KARL A. WITTFOGEL, *Oriental Despotism*

In 1957, Karl Wittfogel proposed the idea that power in Asia was derived
from water. Those who could harness the monsoon rains and transform
them from destructive forces to beneficial assets to aid society became its
rulers. Although some aspects of Wittfogel's "hydraulic state" thesis have
since been called into question, the general contours of his idea remain
powerful and intuitive. Certainly, using natural disasters, particularly water-
borne ones, as a standard against which to measure a ruler's effectiveness
has a long historical pedigree in China. Over the centuries, flooding was a
key indicator that the emperor had squandered the Mandate of Heaven.
In 1938, Generalissimo Chiang Kai-shek lost the hearts and minds of many
Chinese when, without warning local residents, he breached dams along
the Yellow River to halt the invading Japanese army, leaving hundreds of
thousands dead and another million homeless. Even as recently as 1975,
when almost a quarter of a million people were killed after the Shimantan
and Banqiao dams burst, many believed it to foreshadow, like the Tangshan
earthquake a year later, the end of Chairman Mao's rule (he died in September 1976).[1]

---

1. Elizabeth C. Economy, *The River Runs Black: The Environmental Challenge To China's Future*,
Ithaca, NY: Cornell University Press, 2004, 2. 1975 was a bad year for dams in China: the dam
at Zhumadian in Henan province also collapsed after sixty-two reservoirs burst following a
typhoon, killing twenty-six thousand people and "wip[ing] Daowencheng Commune immediately off the map." Chan Siu-sin, "Lifting of Secrecy Veil Sheds Light on Worst Dam Tragedy,"
South China Morning Post, October 2, 2005.

But today's China is no longer the domain of the Son of Heaven, nor the revolutionary experiment of Mao Zedong. It is an increasingly market-driven, decentralized, and politically heterogeneous developing state. In recent years, the control and management of water has slowly transformed from an unquestioned economic imperative to a lightning rod of bureaucratic infighting, societal opposition, and even open protest.

The conventional wisdom, based in no small part on the widespread coverage of the Three Gorges Dam, holds that once a decision is made in Beijing on a project of such magnitude, it will inexorably move forward. Of course, there might be some local resistance or inertia within the implementation stages, but the notion of reversing the decision or even of delaying the project for several years solely as a result of opposition at the local level would be unthinkable. In 1993, this would have been an accurate assessment. Today, it is not.

Consider the following. In the early 1990s, some 179 people were known to have been imprisoned for their vocal opposition to the Three Gorges Dam Project. A book of critical essays on the subject was banned in China and its author, Dai Qing, forbidden to publish or to speak in public.[2] Since 2003, editors and journalists from traditional state-run media mouthpieces as well as more independent newspapers and magazines have written scores of articles critical of contemporary hydropower construction programs and have become important movers and shakers behind the scenes in China's nascent Green movement. They were instrumental in generating the momentum necessary to reverse Beijing's decision to build the Yangliuhu hydropower project in Dujiangyan, Sichuan province. They received no jail time and continue to publish articles and sponsor research critical of Chinese hydropower stations and dam-building. The book *Scientific Valley Development,* a compendium of dam-related issues that tackles some of the most intractable hydropower problems, has become nothing less than a manifesto for hydropower opponents; and although far from a bestseller, it is readily available.[3] In Yunnan province, a critical mass of nongovernmental organizations (NGOs) has emerged with the goal of keeping people in the countryside informed about the pros and cons of dam construction; these NGOs also seek to circumvent the near-total media blackout sought by

2. Dai Qing, *Yangtze! Yangtze!* Toronto, Canada: Earthscan, 1994. Dai was also imprisoned for nearly a year following her involvement in the June 1989 protests. One could make the case that her treatment was in fact relatively lenient—this despite ten months of solitary confinement in Beijing's Qingcheng prison—reportedly due to her considerable family connections within the Chinese Communist Party.

3. *Kexue fazhan guan yu jianghe kaifa* [*Scientific Valley Development*], Beijing, China: Huaxing chubanshe, 2005.

local cadres who stand to gain financially from hydropower management.[4] Even among those technical specialists responsible for the planning and construction of the Three Gorges Dam Project, some have begun to come forward in late 2007 to warn of potential problems once the project is completed in 2009.

These cases suggest a sea change in the structure and process of hydropower politics just in the last decade or so. The contemporary politics of hydropower have provided an unprecedented degree of political pluralism to the Chinese policy process, in which government agencies in opposition to these hydropower projects seamlessly ally themselves with NGOs and, more important, with the third and fourth estates, the public and the press, respectively. Keeping these developments in mind, what if we turn Wittfogel on his head and analyze a "hydraulic society" in tandem with an increasingly complex, diverse, and far less top-heavy hydraulic state? Doing so allows us to examine one of the most fascinating developments in state-society relations unfolding today.

I examine these developments in three case studies—one in which the opposition failed, one in which the opposition succeeded, and one in which the outcome is more ambiguous. In the case of Pubugou, Sichuan province, up to one hundred thousand protesters occupied the dam site and clashed with police; even President Hu Jintao and Premier Wen Jiabao had to weigh in on the subject—sympathetically to the protesters, at that. Yet the protests failed to stop the dam and had the effect only of delaying the beginning of its construction. In the case of Dujiangyan, also in Sichuan province, the outcome was exactly the opposite. Protests at the local level led directly to a reversal of the policy decision at the provincial and national levels to build a hydropower dam at the Yangliuhu site of Dujiangyan. In the final case, the Nu River controversy, the outcome remains uncertain, but the opposition has successfully delayed the project by more than three years; the struggle is not relegated to the corridors and back rooms of the Great Hall of the

---

4. Interview 05KM03A, April 28, 2005. Because this book relies in part on extensive fieldwork with sources that wish to remain anonymous, I indicate interviews by code. The first two digits indicate the year, and the middle letters indicate the location ("BJ" for Beijing, "CD" for Chengdu, "FG" for Fugong, "GS" for Gongshan, "GY" for Guiyang, "GZ" for Guangzhou, "HTX" for the Tiger Leaping Gorge (*Hutiaoxia*) area, "KM" for Kunming, "LJ" for Lijiang, "SG" for Shigu, and "ZQ" for Zhaoqing). The last two digits indicate the overall interview sequence in a given locale. Finally, if there was more than one interview with a particular interviewee, the number of the interview is indicated by the letters A, B, C, etc. at the end of the code. The interview code at the beginning of this citation indicates that it is the first meeting with my third set of informants in Kunming in 2005. The lone exception has to do with the Pubugou case in chapter 3. Because of the sensitive nature of that case, I refer to them all simply as "Pubugou Interviewees," giving the date but not the location of the interview.

People or the leadership compound in Zhongnanhai but instead is being fought in the very public arena afforded by an increasingly liberal media.

The Argument in this Book

There are several arguments that can be put forward to explain such policy outcomes in authoritarian China. But on closer inspection, each falls short in doing so. The first has to do with elite politics in the form of a change in policy resulting from power politics at the higher reaches of the political system. In the cases that follow, elites at the national and local levels do enter into the fray—but often in response to events that have occurred independent of their initial decisions.[5]

A second explanation appeals to changes in the broad ideological "line" of the dominant political party. This may explain a broad policy reversal, but such an argument cannot by itself explain variation within a given policy area. A far better explanation would illustrate why policy reversal occurs in some cases and not in others, while less extreme degrees of policy change occur in yet other circumstances.

A third explanation relies on a cost-benefit calculus. But such a discussion traditionally takes place during the policymaking stage. In each of these three cases, the debate over the hydropower project extended beyond the policymaking stage or arose as a response to the absence of a robust cost-benefit discussion within the government. This suggests that politics drives this process, not market efficiencies. Indeed, when a cost-benefit discourse takes place, it often becomes so politicized that it loses all utility as a bona fide analytical explanation.

A final argument has to do with bureaucratic politics, as expounded in Lieberthal and Oksenberg's seminal work on policymaking and implementation in China.[6] Arguably the most durable framework through which to

5. Although one of the assumptions that guides China's economic development is that society will be better off as it enjoys increasing economic benefits, the reality is often far more complicated. Even though there is a temptation to do so, it is difficult to generalize about elite preferences. President Hu Jintao has recently lavished praise on one of the huge hydropower companies, Huaneng, but in the case of the Nu River Project, Premier Wen Jiabao sided with the State Environmental Protection Administration (SEPA) *against* Huaneng. Where societal preferences bump up against economic development—a prime example being protests by workers laid off because of the need to close down China's moribund state-owned enterprises—political concerns over social stability almost always trump other interests, giving the lie to the official line. Some have also suggested that Beijing sees regulation of its commercial sector and the demand for environmental-friendly technologies as a macroeconomic lever to cool down an overheating economy.

6. Kenneth Lieberthal and Michel Oksenberg, *Policy Making in China: Leaders, Structures, and Processes*, Princeton, NJ: Princeton University Press, 1988.

study Chinese politics is the notion of "fragmented authoritarianism" (FA), in which policy made at the center becomes increasingly malleable to the parochial organizational and political goals of the various agencies and regions charged with enforcing that policy. Policy outcomes result from incorporating the interests of the implementation agencies into the substance of the policy itself. The result is that policy outcomes are often at a considerable variance with the initial goals of the policymakers at the top. FA explains incremental change via bureaucratic bargaining and was derived from three cases within the energy sector of China, one of which, not incidentally, was based in part on hydropower, specifically the Three Gorges Project.

But the FA framework, derived in the mid-1980s, could not possibly take into account the significant changes in the political process since then, particularly the increasing importance of nongovernmental actors and the media.[7] Since the FA framework is derived in part from the policymaking process surrounding the Three Gorges, if the political practices involved in dam-building change—as they have over the past fifteen years—it becomes necessary to revisit and update the FA framework itself.[8]

In this book, I argue that the FA framework remains applicable to the policymaking process in China today, albeit with some important revisions. The politics of hydropower have become increasingly pluralized, as actors and groups that were excluded from the policymaking process in the past are becoming viable players. Political systems, whether democratic or authoritarian, contain a significant degree of fragmentation and agency slack. Whether a result of institutions that have not adapted sufficiently to rapid socioeconomic change, the aggressive lobbying of corporations or interest groups, or changing expectations of the citizenry, fissures emerge within which the state finds it difficult to control the policy process. As has been amply demonstrated in the literature on American politics, these spaces are fertile ground for policy change—if the right set of elements is in place.

Since the 1980s and early 1990s, the political process in China, while no less fragmented, has become somewhat less authoritarian: it has seen the sphere of political conflict increase, allowing actors hitherto relegated to passive recipients of policy outputs to become influential players within the

7. Moreover, given the unique aspect of the Three Gorges Project as the first of the few dam projects that span several provinces, Lieberthal and Oksenberg's analysis may overstate the importance of geographical negotiations in hydropower policy. Either way, this is the one dimension where their analysis and mine are not entirely comparable. In other words, the inter-province bargaining, a hallmark of their analysis, is largely absent here.

8. The extension and updating of the fragmented authoritarianism framework is the subject of chapter 6.

policy process. The roots for this idea of the "expanding of the sphere" go back as far as the early scholarship of Key and Schattschneider, who argued that policy outcomes can be changed by altering the scope of conflict over the issue in question.[9]

In this book, I extend and deepen the FA framework, updating it to capture this dynamic of policy change. To do this, I draw on several strands of the public policy literature in American politics, most significantly three critical dimensions in which the sphere of political conflict and, ultimately, policy change is expanded: policy entrepreneurs, issue framing, and broad support for policy change. I explore each of these interrelated dimensions, separating them for conceptual clarity before bringing them together again to illustrate how the FA framework has evolved from the 1980s to the present day within the policy area of hydropower in China.

Policy Entrepreneurs

John Kingdon defines policy entrepreneurs as "advocates for proposals or for the prominence of an idea" and describes them in the following way:

> These entrepreneurs...could be in and out of government, in elected or appointed positions, in interest groups or research organizations. But their defining characteristic...is their willingness to invest their resources—time, energy, reputation, and sometimes money—in the hope of a future return...[including] in the form of policies of which they approve.[10]

Often, policy entrepreneurs bide their time until chance opportunities arise, perhaps indicating that they are not simply providing a solution in response to an existing problem but waiting for the appropriate problem to arise and gain salience so that they can plug in their already well-developed solutions.[11] As such windows of opportunity emerge, policy entrepreneurs can invest the resources at their disposal to advocate their ideas.[12]

9. V. O. Key, *Southern Politics*, New York: Vintage Books, 1949; and E. E. Schattschneider, *The Semi-Sovereign People*, New York: Holt, Rinehart, and Winston, 1960.

10. John W. Kingdon, *Agendas, Alternatives, and Public Policies*, 2nd ed., New York: Harper-Collins, 1995, 122–23.

11. See, among others, ibid.; Michael D. Cohen, James G. March, and Johan P. Olsen, "A Garbage Can Model of Organizational Choice," *Administrative Science Quarterly* 17 (1972): 1–25; and John C. Campbell, *How Policies Change: The Japanese Government and the Aging Society*, Princeton, NJ: Princeton University Press, 1992.

12. Kingdon, *Agendas, Alternatives, and Public Policies*, especially chapter 8. See also, in different contexts, Campbell, *How Policies Change;* Lynn Kamenitsa, "The Process of Political Marginalization," *Comparative Politics* 30 (April 1998): 313–33; Herbert Kitschelt, "Political Opportunity

One thing that policy entrepreneurs do is interpret events using often existing ideas in a new way, frequently with the goal of convincing potential
supporters. This can be done through "articulation" and "amplification." By
articulating how an issue is described, entrepreneurs link up and assemble
events in order to establish a natural and persuasive narrative, offering a fresh,
alternative perspective on the issue in question.[13] So entrepreneurs pick symbols that can be packaged in such a way that they offer an alternative perspective by which to understand and appreciate events, objects, and situations.[14]

In addition to articulating the issue, policy entrepreneurs amplify the
issue by boiling down the core components of the narrative in order to carry
the frame from one set of individuals to another. Catchphrases or slogans
are popular ways of amplifying the frames for these issues, whether via bum-
per stickers in the United States, chanting in the streets of Paris or Seoul, or
media headlines in China. Other methods include deliberate references to
historical antecedents, metaphors and analogies, and images.[15]

Although policy entrepreneurs are often associated with democratic systems, particularly the United States, fragmented authoritarian systems like
that of China also provide a fertile context for policy entrepreneurs to exist,
and even flourish.[16] The key for them is to occupy spaces from which they
can articulate and amplify their issue in ways that engage the political process rather than existing outside of and in direct opposition to it. Indeed, a
number of policy entrepreneurs in China are veterans of the 1989 protests

and Political Protest: Anti-Nuclear Movements in Four Democracies," *British Journal of Political
Science* 16 (January 1986): 57–85; Doug McAdam, *Political Process and the Development of Black
Insurgency, 1930–1970*, 2nd ed., Chicago: University of Chicago Press, 1999; Sidney Tarrow,
Struggle, Politics, and Reform, Ithaca, NY: Cornell University Press, 1989; and Martha Finnemore
and Kathryn Sikkink, "International Norm Dynamics and Political Change," *International Organization* 52, no. 4 (Autumn 1998): 887–917. I should note here that even though I cite some
pieces from the social movement literature, the arguments I make in this chapter, particularly
the novelty of drawing from American politics to explain policy processes in authoritarian regimes, refer to the public policy literature in American politics; this conceptual link was made
long ago in the social movement literature.

    13. John A. Noakes and Hank Johnson, "Frames of Protest: A Road Map to a Perspective,"
in *Frames of Protest: Social Movements and the Framing Perspective*, ed. Johnson and Noakes, Lan-
ham, MD: Rowman and Littlefield, 8.

    14. Ibid. See also William A. Gamson, "Political Discourse and Collective Action," *Interna-
tional Journal of Social Movements, Conflict, and Change* 1 (1988): 219–44.

    15. Hanspeter Kriesi and Dominique Wistler, "The Impact of Social Movements on Politi-
cal Institutions: A Comparison of the Introduction of Direct Legislation in Switzerland and the
United States," in *How Social Movements Matter*, ed. Marco Giugni, Doug McAdam, and Charles
Tilly, Minneapolis: University of Minnesota Press, 2002, 42–65.

    16. James L. True, Bryan D. Jones, and Frank R. Baumgartner, "Punctuated-Equilibrium
Theory: Explaining Stability and Change in American Policymaking," in *Theories of the Policy
Process: Theoretical Lenses on Public Policy*, ed. Paul A. Sabatier, Boulder, CO: Westview Press,
1999, 102.

and have learned their political lessons well. They recognize that the only way in which such alternative framing of issues is likely to withstand swift retaliation by the state is by pushing the envelope of, but remaining within, the formal and informal rules of political discourse.

Disgruntled Officials

There are three types of policy entrepreneurs in China that figure prominently in the analysis to follow: disgruntled officials, nongovernmental organizations, and the media. The first are officials within Chinese government agencies opposed to a given policy, often because of their official organizational mandates. These units are able to voice their opposition in part because their policy portfolios give them a degree of political cover. Conversely, by refraining from pursuing their organizational mandates, these units run the risk of being seen as weak or even irrelevant, a potentially deadly label in the current era of administrative downsizing and bureaucratic fat-cutting.[17]

The State Environmental Protection Administration (SEPA) and its local environmental protection bureaus (EPBs) are often on the front lines of the opposition. This bureaucracy (*huanbao*) is charged with managing environmental protection by enforcing the relevant laws and regulations. SEPA is a ministry-level bureau (it was upgraded in the late 1990s), and one of its current vice directors, the maverick Pan Yue, is something of a media darling due to his often outspoken comments on environmental issues in China. And Pan's boss, SEPA director Zhou Shengxian, appears to be every bit as much an activist as his deputy. This combination of increased authority and exposure has allowed this bureaucracy to be more aggressive in pursuing its institutional mandate.

Another set of actors includes the officials within the cultural relics (*wenwu*) offices embedded within the Ministry of Culture bureaucracy. This office is charged with the protection of cultural artifacts threatened by the construction of dams, hydropower, or other projects. The cultural relics office draws its mandate from the Cultural Relics Law of 2002.[18] Because of China's rich cultural heritage, in some cases this group has a degree of

---

17. Andrew C. Mertha, "Shifting Legal and Administrative Goalposts: Chinese Bureaucracies, Foreign Actors, and the Evolution of China's Anti-Counterfeiting Regime," in *Engaging the Law in China: State, Society, and Possibilities for Justice*, ed. Neil J. Diamant, Stanley B. Lubman, and Kevin J. O'Brien, Stanford, CA: Stanford University Press, 2005, 161–92, and "Policy Enforcement Markets: How Bureaucratic Redundancy Contributes to Effective Intellectual Property Implementation in China," *Comparative Politics* 38, no. 3 (April 2006): 295–316.

18. Available at www.gov.cn/english/laws/2005-10/09/content_75322.htm, accessed May 12, 2007.

leverage not necessarily available to the EPA. On the other hand, the ability of a locale to leverage cultural concerns is partly a function of the degree to which it possesses such an endowment.

Bureaucrats within local World Heritage (*shijie yichan*) offices are another source of policy entrepreneurship for the cases in this book. There are currently some three dozen World Heritage sites in China. The relationship of these officials to the United Nations Educational, Scientific, and Cultural Organization (UNESCO) office in Beijing is somewhat ambiguous. They are not local representatives of UNESCO; rather, they are adjunct offices of local governments. These logic-defying governmentally organized nongovernmental organizations (GONGOs) can be important players in those increasingly pluralistic political and policy process.

Other officials unlikely to support such projects include those in the seismology (*dizhen*) and construction (*jianshe*) bureaucracies. The former are weary of such projects because of the increased mass of the water retained in reservoirs on potentially unstable land, particularly over fault lines.[19] The construction bureaus generally oppose dam and hydropower construction because the bureaus are ineligible for the actual construction of these projects. As far as the communications (*jiaotong*) bureaucracy is concerned, the opportunity cost of roads and other infrastructure that might otherwise have been built in the flooded areas helps tilt the communications bureaucracy against dam projects, all things being equal. These units often offer less pronounced opposition, relative to the EPA and the cultural relics bureaus. Nonetheless, when every potential ally matters, these units should not be counted out—nor are they.

Other political fissures are geographical, arising from the spatial fragmentation of China's decentralized political system. For example, although local governments are the first to see the economic benefits of hydropower projects and have several ways—both above board and under the table—that they can benefit from the sometimes considerable revenue streams associated with them, in cases where such projects actually threaten the economic well-being of a locality—for example, if they cut into tourism revenues or threaten local hydropower and other infrastructure projects—they can be formidable opponents of such projects. This opens up a whole new set of cleavages, as is amply documented by the extensive literature on Center-local relations in China, to be potentially exploited by policy entrepreneurs.[20]

---

19. Cathy Shufro, "Damming Tiger Gorge," *E Magazine*, January 1, 2005, and Shi Jiangtao, "Landslides Pose a Bigger Threat," *South China Morning Post*, January 4, 2005.

20. See, among others, Michel Oksenberg and James Tong, "The Evolution of Central-Provincial Fiscal Relations in China, 1971–1984: The Formal System," *China Quarterly* 125

The Media

A second major category of policy entrepreneurs is comprised of journalists and editors in an increasingly liberal media environment. Although it is important to avoid overstating the growing parameters of acceptable discourse in the Chinese media, newspapers, magazines, and television broadcasts have provided a platform for journalists to pursue stories that match their own increasingly progressive interests and agendas. This, in turn, has been reinforced by a Chinese media increasingly required to generate its own budgetary revenue. As a result, it must rely on advertisers who will not pay for ad space if people do not consume said media. To ensure that people do so, there has been a dramatic increase in the proportion of tabloid journalism stories that, in addition to racy sex stories, cover government injustice, civil protest, and the like.[21] In some cases, this new aggressiveness on the part of the Chinese media outlets has led to some unintended humorous consequences, as when the *Beijing Evening News* reprinted as a genuine news story a May 23, 2002, article appearing in *The Onion* in which Congress was satirically depicted as evaluating competing bids to relocate based on which city could build the best new Capitol building (pictures of which looked remarkably like a sports stadium).[22]

Moreover, this is not limited to such maverick publications as *Southern Weekend* (*Nanfang Zhoumou*) or *Finance* (*Caijing*); it is also the case in the more traditional bastions of official propaganda such as the *China Youth Daily* (*Zhongguo Qingnian Bao*) and other mainstream news outlets. The manager of danwei.org, a website that tracks media developments in China, Jeremy Goldcorn, even refers to *Xinhua* (New China News Agency) as "Sin*hua*—the state-run lads' mag."[23] This explains in part the expanding footprint of the media outlets in the policy and political process in China today. This growing aggressiveness is what helps translate dissatisfaction with a given policy into the articulation of an opposing viewpoint that can mobilize like-minded individuals and organizations into political action. A growing source of the

---

(March 1991): 1–32; Jean Oi, "Fiscal Reform and the Economic Foundations of Local State Corporatism in China," *World Politics* 45, no. 1 (October 1992): 99–126; Dali Yang, "Reform and the Restructuring of Central-Local Relations," in *China Deconstructs: Politics, Trade, and Regionalism*, ed. David Goodman and Gerald Segal, London: Routledge, 1994; and Le-Yin Zhang, "Chinese Central-Provincial Fiscal Relationships, Budgetary Decline, and the Impact of the 1994 Fiscal Reform: An Evaluation," *China Quarterly* 157 (March 1999): 115–41.

21. Daniel C. Lynch, *After the Propaganda State: Media, Politics, and "Thought Work" in Reformed China*, Stanford, CA: Stanford University Press, 1999.

22. "China Paper Bites on Onion Gag," Reuters, June 7, 2002.

23. Peter Goff, "Now It's Sinhua, the State-Run Lads' Mag," *South China Morning Post*, October 12, 2005.

media's power is the close relationship it shares with many Chinese nongovernmental organizations.

Nongovernmental Organizations

NGOs are a critical set of actors that define the contours of policy entrepreneurship in China. There are estimated to be as many as 1.5 to 2 million NGOs in China today. Rather than dismissing them as merely symbolic or politically supine in an authoritarian state like China, it is far more instructive to analyze their actual role in the policy process. Michael Büsgen cogently argues that NGOs in China are different from those that helped bring about regime change in the Soviet Union and Eastern Europe because they must work with the Leninist party state and must be equally sensitive to political continuity if they wish to bring about change within their particular policy areas.[24]

In several different experiments, the Chinese government has attempted to enhance state capacity by entering into partnerships with commercial and other units that do not have an overtly political function as a way to improve the quality of governance in China. These relationships have been forged to meet some diverse challenges, from regulating the marketplace to providing social services to empowering environmental watchdogs to help implement existing laws that have largely gone unenforced.

Other NGOs were established to help "wean" (literally, *duannai*) officials from the government payrolls and to absorb the layoffs and retirements that resulted from the 1998 governmental restructuring.[25] Some GONGOs, such as local World Heritage offices, have developed keen political antennae as well as formidable political teeth despite their formal administrative weakness. Still other NGOs have emerged on university campuses in much the same way that their counterparts have emerged in the West. Moreover, official figures probably underestimate the actual number of NGOs because of the cumbersome process of finding a political sponsor or "host unit," with many NGOs operating in a "not-quite-official-yet" capacity as they await official approval.

One thing that accounts for the successes of NGOs in Chinese politics is that an inordinate number of their officers and staff members were trained

---

24. Michael Büsgen, *NGOs and the Search for Chinese Civil Society: Environmental Non-Governmental Organisations in the Nujiang Campaign*, Master's thesis, Institute for Social Studies, Graduate School of Development Studies, The Hague, Netherlands, 2005, 2–3.

25. For a good overview of the institutional changes stemming from the Ninth National People's Congress, see Dali Yang, *Remaking the Chinese Leviathan: Market Transition and the Politics of Governance in China*, Stanford, CA: Stanford University Press, 2004.

as journalists or editors, giving them especially close access to the media. This almost seamless interface between the two groups allows Chinese NGOs to play a more significant role in the political process than might otherwise be the case.

Yet policy entrepreneurship is far from a sufficient condition for opposition-led policy change. In political discourse, issue framing looms very large, and the establishment of the frame itself is critical to the overall success or failure in mobilizing the opposition.

Issue Framing

In their landmark book *Agendas and Instability in American Politics,* Baumgartner and Jones argue that

> As a consequence of the dynamics of the allocation of attention, the partial equilibria of policy monopolies tend to be temporary. Moreover, they tend to be disrupted turbulently rather than gradually...existing policy subsystems are overwhelmed by a flood of new participants or by dramatic new policy proposals. Over the long run, open, democratic political systems are characterized both by policy monopolies, as the political system struggles with its limited capacity to process numerous issues simultaneously, and by turbulent disruptions, as attention is directed at the issue again. That is, democratic political systems are composed of punctuated partial equilibria.[26]

Notwithstanding the suggestion of Baumgartner and Jones that punctuated equilibrium is an outgrowth of a particular political regime type, as a descriptive heuristic it is particularly apt in illustrating how China's fragmented authoritarian system has become increasingly pluralized in recent years. A key element in this process, and one that resembles their notion of punctuation of partial policy equilibria, is issue framing. Policy entrepreneurs shape the contours of political discourse and thus mobilize allies toward the goal of policy change by "organizing information in a manner that conforms to the structure of a good story."[27]

Even if the opposition fails to alter the policy at hand, issue framing in authoritarian regimes often leads to a policymaking process that is far more pluralistic in scope than was initially anticipated by those in power. Unlike the case of Eastern Europe at the time of the fall of communism in the 1989–1991 period, framing as explored in this book does not signal a

26. Frank R. Baumgartner and Bryan D. Jones, *Agendas and Instability in American Politics,* Chicago: University of Chicago Press, 1993, 20–21.

27. Adam J. Berinsky and Donald R. Kinder, "Making Sense of Issues through Media Frames: Understanding the Kosovo Crisis," *Journal of Politics* 68, no. 3 (August 2006): 640.

prelude to change in regime type. Rather, issue framing is quite compatible with the evolving, nondemocratic but increasingly pluralistic political processes of authoritarian regimes, provided the goal of critics is not fundamental regime change. Thus, rather than a certain regime type, the important precondition is the existence of spaces or fissures in the political and institutional landscape where incremental changes can enter, gestate, and eventually overwhelm an entrenched policy. To paraphrase Huntington, the *type* of government is less important than the *degree* of government in explaining such outcomes.[28]

The concept of framing suggests a few dimensions not captured in the current literature on policymaking in China, particularly with regard to the ability of the media to frame coverage of a particular story and to mobilize support.[29] Most fundamentally, it suggests that significant clusters of media outlets have been able to transform their former function as enablers of state policy to a role typically associated with the fourth estate. Second, it demonstrates the growing influence of NGOs within the policy process in China through the "moonlighting" by NGO staffers as journalists and editors ("wearing different hats"—*dai jiding maozi*—in Chinese parlance). Finally, it lays bare the intimate contact between these media outlets and government leaders as political allies, both in Beijing and in the localities.

The success of these "alternative" media frames is due in part to the decline in the potency of official "state framing," a mainstay of authoritarian political regimes like that of China.[30] The infallibility of Mao Zedong Thought expired before Mao himself did, and the Reform Era has been marked by a degree of skepticism about official ideological exhortations. As a result, "official" state framing increasingly has been vulnerable to tinkering from outside the propaganda apparatus. Some types of framing appear to be quite strong—the shift of Falun Gong from a somewhat suspect homegrown spiritual society to its current articulation as a "poisonous cult"—and

28. Samuel P. Huntington, *Political Order in Changing Societies*, New Haven: Yale University Press, 1968, 1.

29. This is not to say that the concept of framing itself has not been applied to China. See, for example, William Hurst, "Mass Frames and Worker Protest," and Chen Feng, "Framing Contention: The Role of Worker Leaders in Factory Based Resistance," in *Popular Contention in China*, ed. Kevin J. O'Brien, forthcoming. However, at present, issue framing in China has largely focused on worker protest and unrest. See also Patricia M. Thornton, "Framing Dissent in Contemporary China: Irony, Ambiguity, and Metonymy," *China Quarterly* 171 (September 2002): 661–81.

30. John A. Noakes, "Official Frames in Social Movement Theory: The FBI, HUAC, and the Communist Threat in Hollywood," in *Frames of Protest*, ed. Johnson and Noakes, 89–111.

are backed up by the state's coercive apparatus.[31] Similarly, one is likely to find very little deviation from the government "line" among Mainland Chinese on hot-button issues like the "Three T's": Taiwan, Tibet, and Tiananmen. However, other subjects, such as the rush to welcome capitalists into the Chinese Communist Party (CCP) under the aegis of the "Three Represents" (*sange daibiao*), are quite capable of eliciting scorn in citizens and even, albeit behind closed doors, in some officials. Nevertheless, official framing in China is anything but a figment of the imagination; it can be potent and persuasive, as in Beijing's casting of dams as gateways to development and modernization, reminiscent of Nehru's invocation of dams as "temples of modern India."[32]

Far more significant than the decline of official state framing, however, is the rise of unofficial, alternative issue framing in China. Although policy entrepreneurs deliberately choose from a wide array of symbols to construct their issue frames, these frames must be meaningful to the people whom they wish to mobilize for their cause. Moreover, the process is not a fishing expedition or an arbitrary mining of such symbols but rather is calculated to employ strategically those elements that will resonate as robustly as possible and thus draw the greatest number of potential recruits. At the same time, entrepreneurs must be careful to avoid overstating their case—for example, drawing on such universal values as "social justice"—without demonstrating how such appeals matter in a very personal way for their potential allies.[33]

Frames have to appear natural, logical, and recognizable, and finding the right one can be tricky: "Some metaphors soar, others fall flat; some visual images linger in the mind, others are quickly forgotten."[34] There are several dimensions that improve the likelihood that frames will be successful. First, frames must be "culturally compatible": they must resonate with the general "cultural stock" of society as well as with the specific "cultural tool kit" of the specific target.[35] Frames that succeed in this tap into not just shared everyday experiences but also broader cultural mores. Second, these frames have

31. I happened to be in Beijing during the April and July 1999 protests by the Falun Gong and witnessed the rapid shift in frame, with Falun Gong going from one of many organizations practicing a version of *qigong* (traditional Chinese breathing exercises) to an anti-CCP cult. It was very impressive to witness the speed and thoroughness with which the state could establish this new issue frame.

32. Ken Conca, *Governing Water: Contentious Transnational Politics and Global Institution-Building*, Cambridge, MA: MIT University Press, 2006, 170.

33. Noakes and Johnson, "Frames of Protest," 11.

34. William A. Gamson, *Talking Politics*, New York: Cambridge University Press, 135.

35. Anne Swidler, "Culture in Action: Symbols and Strategies," *American Sociological Review* 51 (April 1986): 273–86.

Table 1.1 Selected bottom-up oppositional campaigns in China

Year	Events
1995–1996	Protecting the Golden snub-nosed monkey
1997	Boycotting disposable chopsticks
1997	Promoting campus recycling
1997	Guarding the wild geese in Purple Bamboo Park, Beijing
1998–1999	Protecting the Tibetan antelope
2000	Earth Day campaign
2000	Protecting the Tibetan antelope (Internet campaign)
2001	Boycotting "wild tortoise" medicinal products
2001–2003	Protecting the Jiangwan wetlands in Shanghai
2002	Protesting the building of an entertainment complex near the suburban Beijing wetlands (Internet campaign)
2003	Fighting SARS
2003	Protecting the Dujiangyan World Heritage site
2005	Yuanmingyuan (summer palace) campaign
2003–2006	Stopping the proposed dam on the Nu River

Source: Guobin Yang, "Is There an Environmental Movement in China? Beware of the 'River of Anger,'" *Asia Program Special Report* 124 (September 2004).

to be internally consistent with the movement's beliefs and goals. Finally, such collective action frames have to demonstrate a requisite degree of relevance to the target audience. The latter must be drawn into the issue and actually care about it. Otherwise, such frames are likely to fail in mobilizing potential adherents outside of the core group of activists.[36]

On the other hand, policy change must be understood by the powers that be as being distinct from actual political change, that is, it must not signal a sharp deviation from the political status quo that might threaten their hold on power. The effectiveness of modest policy goals is underscored by a partial list of successful bottom-up campaigns in China from 1995 to 2006 (see Table 1.1). In none of these cases was there anything approaching a fundamental shift in the structure of the state, only a shift in policy. This, however, does not dilute the importance of policy as distinct from political change:

> Policies are generally more easily altered than the constitutive rules of formal institutions, but they are nevertheless extremely prominent. Policies, grounded in law and backed by the coercive power of the state, signal to actors what has to be done and what cannot be done, and they establish many of the rewards and penalties associated with particular activities.[37]

36. Noakes and Johnson, "Frames of Protest," 15.
37. Paul Pierson, "Path Dependence, Increasing Returns, and the Study of Politics," *American Political Science Review* 94, no. 2 (June 2000): 259.

Recalling Baumgartner and Jones, successful issue frames are those that allow "existing policy subsystems" to become overpowered by the "flood of new participants or by dramatic new policy proposals."[38] I discuss the role of these new participants in the next section.

Coalitions and Broad-Based Support

Policy entrepreneurs provide part of the supply side of the political equation. But in order to be successful, the issue frames they create must link these entrepreneurs to another necessary component of any expansion of the political sphere of conflict: coalitions and broad-based support. Both of these elements are secured by elevating what may otherwise be a local issue to the supra-local level, that is, onto the national or even the international stage.

It is more difficult for political opponents to quash an issue once it extends beyond their local or parochial control, once it garners national-level attention. Moreover, broad-based support can itself become self-sustaining and even self-generating as many potential recruits become aware, in some cases for the first time, that they are not alone in possessing some policy preference over the issue in question. The advocacy coalition framework describes a wide range of participants who can work together to bring about change.[39] These coalitions include not only the traditional iron triangle of administrative agencies, legislators, and interest groups but also journalists, researchers who are receptive to new ideas, and actors at all levels of government affecting the policy in question.[40]

The processes examined in this book focus on traditional Chinese political organizations and newly significant political actors working together. Interest groups outside of the official state apparatus liaise with government agencies, sometimes through professional networks. What is necessary is simply the existence of a space in which groups can function without the threat of being shut down by the authorities. It also requires that the political system be susceptible to the pressures that emerge to some degree, no matter how small. Most authoritarian regimes, regardless of the image they deliberately project, do possess such characteristics.

---

38. Baumgartner and Jones, *Agendas and Instability in American Politics*, 20–21.

39. Paul Sabatier and Hank Jenkins-Smith, *Policy Change and Learning: An Advocacy Coalition Approach*, Boulder, CO: Westview Press, 1993, and "The Advocacy Coalition Framework," in *Theories of the Policy Process*, ed. Sabatier, Boulder, CO: Westview Press, 1993, 117–66.

40. Sabatier and Jenkins-Smith, "The Advocacy Coalition Framework," 119. On the political role of journalists, see Stanley Rothman and S. Robert Lichter, "Elite Ideology and Risk Perception in Nuclear Energy Policy," *American Political Science Review* 81 (June 1987): 383–404;

Of course, *coalition* is, politically speaking, a highly charged term in China. But these combinations of interest groups are definitely coalitions, as they "share a set of normative and causal beliefs" and "engage in a nontrivial degree of coordinated activity over time."[41] Moreover, they are relatively nascent groups and have not had the luxury of working together for a decade or more. Rather, they conform to the hypothesis that "actors who share policy core beliefs are more likely to engage in short-term coordination if they view their opponents as (a) very powerful and (b) very likely to impose substantial costs upon them if victorious."[42]

Any issue that rises up to the national level or beyond has a chance of mobilizing an even greater segment of the population that, while not as actively engaged in the issue as policy entrepreneurs or active coalition members, can nevertheless demonstrate their views on the policy in question. Mass groups have played an increasing role in the process of dam politics in China on several different dimensions. The first is that the segment of the population targeted for resettlement (*yimin*) may receive what it considers inadequate compensation.[43] This demographic can easily be tapped for protest and has been an important part of the evolution of the politics of dam building in the past decade or so. It is estimated that in 2003, three million people were involved in 58,000 protests, up 15 percent from 2002.[44] In 2004, that number jumped to 72,000, and in 2005, it rose again to 87,000 (50,000 of which were environmental protests).[45] Peasants, once seen as reliably fatalistic by government authorities, are becoming far less so.[46]

Another dimension is those elements of society with "green" or other "postmaterialist" tendencies that get their information from the media.[47]

---

and Shanto Iyengar, *Is Anyone Responsible? How Television Frames Political Issues,* Chicago: University of Chicago Press, 1991.

41. Sabatier and Jenkins-Smith, "The Advocacy Coalition Framework," 120.

42. Ibid., 140.

43. Unlike the conventional wisdom surrounding the Three Gorges case, much of the protests surrounding the people to be resettled (*yimin*) revolved around economic issues, that is, whether or not the affected parties would get their fair share of the pie. From my interviews, such mobilization did not appear to center on environmental or cultural issues, as implied in much of the reporting on the Three Gorges controversy.

44. Kathy Chen, "Chinese Protests Grow More Frequent, Violent," *Asian Wall Street Journal,* November 5, 2004.

45. Li Fangchao, "Environment Issues to Be Addressed More Urgently," *China Daily,* May 4, 2006.

46. Josephine Ma, "Mouse That Roared over Tiger Leaping Gorge," *South China Morning Post,* November 19, 2004; and Christopher Bodeen, "Report: One Killed in Protest over Dam in Western China," *Associated Press,* November 1, 2004.

47. See, among others, Ronald G. Inglehart, *The Silent Revolution: Changing Values and Political Styles in Advanced Industrial Society,* Princeton, NJ: Princeton University Press, 1977.

Finally, if an issue hits people in the "gut," the potential parameters of the sphere increase exponentially, encompassing even those otherwise not inclined toward political participation:

> The mobilization of the apathetic provides the key to linking the partial equilibria of policy subsystems in American politics to the broader forces of governance. As different groups become active on a given issue, partial equilibria of preferences are altered quickly from one point to another. Apathy is the key variable in politics. Some seek to promote it, others to fight it. Depending on the degree of apathy that prevails, different groups will see their views adopted as the majority view. As the level of apathy changes, so do majority opinions.[48]

In sum, policy entrepreneurs and larger coalitions with political allies and mass constituencies are linked substantively and conceptually by the ways in which the former articulate the policy through issue framing in order to mobilize the latter.

Variation in Outcomes

The foregoing presents a broadening of the various strands of public policy literature derived from American politics. Although the literature on social movements has been extended beyond particular regime types (i.e., democracies) to other political contexts, the public policy literature has been somewhat less successful in this regard. Therefore, for the scholar otherwise uninterested in China or in hydropower, the analysis to follow provides a rich example of how ideas with their roots in the context of American politics can account for political and policy change in an authoritarian state such as China.

A fragmented political system provides policy entrepreneurs a key resource necessary to compete politically within the policy process, that is simply the "spaces" necessary for them to exist without being snuffed out by the coercive apparatus of the state. In fragmented political systems, territorial, jurisdictional, and other political cleavages provide comparatively fertile ground for various contending state interests to push their agendas and to arrive at compromises that better reflect their own parochial or institutional goals. This is well documented in the literature on China's bureaucratic politics.[49] And this is exactly the method employed by the policy

---

48. Baumgartner and Jones, *Agendas and Instability in American Politics*, 21.
49. Lieberthal and Oksenberg, *Policy Making in China;* Susan L. Shirk, *The Political Logic of Economic Reform in China,* Berkeley: University of California Press, 1993; Carol Lee Hamrin, *China and the Challenge of the Future: Changing Political Patterns,* Boulder, CO: Westview Press,

Table 1.2 Schematic of the argument

Policy entrepreneurship	*Dominance of oppositional issue frame*	
	High	*Low*
High	Dujiangyan (chapter 4)	Nu River (chapter 5)
Low	Three Gorges	Pubugou (chapter 3)

entrepreneurs in China. In other words, the political dynamics captured in the fragmented authoritarianism framework provide policy entrepreneurs with a road map, a playbook by which they can pursue their policy goals. They adopt the strategies that traditional institutions have used for decades to pursue their agendas and institutional mandates.

Although it is beyond the scope of this analysis to actually test the claims made herein, Table 1.2 provides a universe of ideal-type outcomes based on the foregoing argument and places the three cases explored in this book within it.[50] In the two upper cells, values for the emergence of policy entrepreneurs are relatively high. There are several conditions that are auspicious for policy entrepreneurs to emerge, and they suggest why and under what conditions policy entrepreneurship is more likely. The first is that although these individuals are increasingly disposed to act strategically in pursuing their policy goals, for most people the initial motivation is based on a genuinely affective, even deeply emotional, relationship with the policy in question. For many of them, these issues have a deep, personal resonance that is important enough for them to expend their limited resources and to undertake risky political action in pursuit of them.

As many policy entrepreneurs described in this book have learned from past failed attempts at political opposition to state policies, these individuals are not at all naïve about the importance of securing some sort of political cover to insulate their activities from the state's ability to close them down.

1990; Carol Lee Hamrin and Zhao Suisheng, eds., *Decision-Making in Deng's China: Perspectives from Insiders*, Armonk, NY: M. E. Sharpe, 1995; Kenneth Lieberthal and David Lampton, eds., *Bureaucracy, Politics, and Decision Making in Post-Mao China*, Berkeley, CA: University of California Press, 1992; and David Lampton, "Water: Challenge to a Fragmented Political System," in *Policy Implementation in China*, ed. David Lampton, Berkeley, CA: University of California Press, 1987, 157–89.

50. The research design of this study is that of a hypothesis-*generating* (not-testing) case study approach. See Arend Lijphart, "Comparative Politics and the Comparative Method," *American Political Science Review* 65, no. 3 (September 1971): 682–93. Moreover, this research design is to test existence/impossibility claims rather than hypotheses. See Robert Pahre, "Formal Theory and Case-Study Methods in EU Studies," *European Union Politics* 6, no. 1

As mentioned previously, some government officials can hide behind the functional responsibilities of their offices. Others have information connections that can insulate them against political sanctions. SEPA's Pan Yue is the son-in-law of Liu Huaqing, former head of the Chinese navy, Politburo Standing Committee member, and member of the Central Military Commission. Activist Liang Congjie, the founder the NGO Green Earth Volunteers, is the grandson of late Qing Dynasty reformer and intellectual Liang Qichao and the son of master architect and urban planner Liang Sicheng, who incurred Mao's wrath when he argued against leveling the traditional neighborhoods of Beijing in order to preserve China's cultural heritage. These connections have no doubt served these individuals well.

Finally, some activists have become such high-profile personalities internationally that it becomes difficult—but by no means impossible—to silence them. These dimensions of policy entrepreneurship in China combine to provide these individuals a requisite degree of credibility.

Other factors that contribute to such credibility include measurable professional expertise, a willingness to take risks, personal charisma, and the ability to forgo such projects that might undermine or dilute that same credibility. Yet another dimension has to do with the nature of the policy as well as the specific characteristics of the cases themselves. This variation in the specific (physical, jurisdictional, geographical) characteristics of the case at hand accounts in part for the variation in the "spaces" necessary for policy entrepreneurs to initiate their activities within the policy process. For instance, if a particular site is located in an area in which there is considerable risk of seismological activity, pollution, loss of cultural relics, and opportunity costs in terms of other infrastructure projects, it has a greater pool of potential policy entrepreneurs from functionally related bureaus (the seismological bureau, the environmental protection administration, cultural relics bureau or world heritage office, the bureau of communications, and so on) as well as outsiders (interested NGOs and journalists). This does not mean that all of these actors will necessarily mobilize, simply that they might.

Variation across the two upper cells has to do with the dominance of the issue frame in question. In the case of Dujiangyan, the subject of chapter 4 (the upper-left cell), the oppositional issue frame achieved an almost insurmountable dominance. As was the case in Dujiangyan, opposition that is framed in terms of cultural heritage issues is extremely effective in

---

(January 2005): 113–46; and Andrew Mertha and Robert Pahre, "Patently Misleading: Partial Implementation and Bargaining Leverage in Sino-American Negotiations on Intellectual Property Rights," *International Organization* 59, no. 3 (Summer 2005): 695–729.

grabbing the attention of a larger potential audience and, moreover, in an affective fashion that transcends other dimensions that might otherwise mitigate the effect of the frame. The outcome is not assured, however: the alternative issue frame of cultural heritage issues did not achieve a similar degree of dominance in the case of the Three Gorges, largely because of the politically induced absence of policy entrepreneurship.[51]

In the Nu River case analyzed in chapter 5 (the upper-right cell), policy entrepreneurship was extremely high, but the oppositional issue frames were not quite as dominating as was the case in Dujiangyan. This was partly because policy entrepreneurs were mobilized and issue frames were invoked on all sides of the issue. The oppositional frame of environmental protection never achieved the degree of salience that the cultural heritage frame evoked at Dujiangyan did. Environmental issues evolved along two trajectories, in an emotive way or along a scientific dimension. The first, the notion of "worshipping nature" (*jingwei daziran*) can easily be derided by opponents that stress scientific inquiry, whether the ends are political or scientific. Insofar as the opposition seeks to meet the scientifically based challenge of the policy proponents, they are handicapped because of the lack of scientific data that is available to the opposition. Such data are jealously guarded by government agencies and the scientific institutes in their employ. Even when environmental concerns combine with more "material" concerns, it is possible to separate the two—to divide and conquer—as such groups, usually differentiated on the basis of a "material" and "purposeful" bifurcation, may not share important core beliefs. One result is that something as seemingly objective as a "scientific" debate over the effects of economic development on the environment quickly becomes politicized, whereby oppositional issue frames compete with, rather than dominate, the official state frame.

The bottom-right cell illustrates the outcome when both policy entrepreneurship and oppositional issue frame dominance are lacking: the maintenance of the status quo with little or no change. In the case of Pubugou, covered in chapter 3, policy entrepreneurship was largely absent once the opposition took the form of large-scale and potentially violent demonstrations. Moreover, the issue frame focused on narrow social justice claims, specifically on the inadequacy of resettlement compensation. Opposition framed around compensation seems to force local government agencies to dig in their heels, scaring off potential allies within these agencies.

51. On the variability of the efficacy of cultural heritage framing in a different context, see Ian Johnson, *Wild Grass: Three Stories of Change in Modern China*, New York: Pantheon, 2004, ch. 2.

Moreover, the resonance of this issue frame among mass publics is also limited. Many people in China face problems of social justice in their daily existence. It is difficult to expect people to be overwhelmingly empathetic toward the hardships faced by others, even those who find themselves in far more difficult situations. And once the issue becomes one of economic gain, it loses much of the goodwill among this potential audience and thus loses its power to attract the broad support necessary for policy change.[52]

Finally, there is the question of what a case in the lower-left cell might look like, that is, when an oppositional issue frame is relatively high but policy entrepreneurship is low. Such a situation is somewhat at odds with the argument I make in this book, which takes it as a given that issue framing represents a conscious and deliberate strategy undertaken by policy entrepreneurs, that such framing does not simply arise out of nowhere. Although there is no such case discussed in this book for this reason, one can tease out an approximation of what such a situation *might* look like by looking at the Three Gorges Dam Project. I refer to this case here with some trepidation insofar as key aspects of it do not match up with the other three cases. It predates the other cases by a good fifteen years, it involved overlapping jurisdictional dimensions not captured in the other three cases, and it represented a mélange of most of the official and oppositional issue frames described in this book. As a result, its usefulness for comparison is extremely limited. Thus, rather than treat it as a separate empirical case, I use it very loosely as a heuristic device, almost as a counterfactual—albeit one inspired by real events.

That said, one might argue that because of the political climate at that time, policy entrepreneurship that existed outside of traditional, official state channels was largely curtailed during the decision-making and initial implementation processes. Nevertheless, domestically and internationally, oppositional issue frames had achieved a degree of dominance through media reports and discourse, to the point that in some circles, it became difficult to understand why the state would undertake such a massive project in the first place.

Among the official players in this policy debate—that is, line ministries and local governments—the result was a political compromise that had significant effects on the contours of the policy that was eventually adopted.[53] Although there was opposition within the National People's Congress, policy entrepreneurs were largely absent; those who did speak out were quickly

52. Pubugou Interviewee, December 12, 2006.
53. See Lieberthal and Oksenberg, *Policy Making in China,* esp. ch. 6.

silenced or anonymously voted against it.[54] Oppositional issue frames did emerge, encompassing cultural heritage, environmental, and resettlement/ compensation issues, but they largely wafted above and outside of the decision-making corridors of power and were mostly irrelevant to the decision to move forward. Once the decision was made, those issue frames never seriously threatened it.

In sum, while policy entrepreneurship appears to be a necessary condition for policy change, it alone does not guarantee that such policy change will occur or, if it does occur, whether it will be permanent or the degree to which it represents a compromise that may cut against some of the core policy goals of the opposition. For victory to be decisive, it is necessary to invoke an issue frame that can overwhelm the official state frame used to legitimize the policy in the first place by expanding the sphere of political conflict through the mobilization of new groups into the policy process.

Broader Generalizability and Significance

Few issues explain the structure and process of politics and policymaking better than the politics over water policy. Because of technical complexity, the functional administrative units involved, the spatial dimensions of waterways that defy administrative and political boundaries, and the substantive issues at stake, the politics over water policy are unique in the degree to which they allow administrative and geographical conflict and debate within the political process, which, rising in sharp relief, allow us a glimpse into a political space that is often opaque at best. This is one of Lieberthal and Oksenberg's central claims, and it remains one of the most powerful dimensions of their conceptual framework. Looking at these processes longitudinally, by combining their study with this one, allows us to observe how the machinery of the state operates and to track changes over time.

The politics of hydropower in China allow us to evaluate other broad contours of the evolving Chinese state. These include center-local relations, the debate over the nature (indeed, existence) of Chinese quasi-federalism, bureaucratic conflict vis-à-vis discrete bureaucracies and broader clusters of functionally-related bureaucracies (*xitong*) and policymaking, to name a few.

---

54. The vote on the Three Gorges project was taken at the NPC on April 3, 1992; 1,767 NPC delegates voted in favor, with an unprecedented 177 votes against and 644 abstentions. See Shi He and Ji Si, "The Comeback of the Three Gorges Dam (1989–1993)," 37.

Comparatively speaking, the domestic processes we are witnessing in China over the contested terrain of hydropower bear an uncanny resemblance to similar processes in other developing states, such as India. Contentious hydropower politics are a global phenomenon. In 1982, 369 people were killed in clashes following their opposition to the Chixoy Dam in Guatemala. Violent state crackdowns were launched against protesters of the Itoiz Dam in the Basque region of Spain. And in Thailand, a village set up by protesters at the site of the Pak Mun Dam was the scene of clashes between protesters and agents of the hydropower company.[55]

But the comparison does not stop there. Conceptually, this analysis demonstrates that positive outcomes, specifically changes in norms and in the political processes governing hydropower policy, are also increasingly similar to those in nondeveloping and non-authoritarian regimes such as those in the United States and Australia.[56] Moreover, these political processes are not only linked to dams, hydropower, and energy, they can also be likened to such disparate historical phenomena as the civil rights movement of the 1950s and 1960s in the United States and the anti-nuclear and environmentalist movements of the 1980s in Europe.

But the issue of hydropower, even when pared down from broad generalizability claims, is nonetheless an extraordinarily important issue in China. Of the forty-five thousand large-scale dams (fifteen meters or higher) in the world, about half are in China. In contrast to trends in the United States to replace larger dams with smaller ones, in China the trajectory over water use seems to be exactly the opposite, illustrated most dramatically by the current South-North Water Transfer Project (*nanshui beidiao*). Other cases include plans for a four-part hydropower project along the Jinsha River in Yunnan at Xiluodu, Xiangjiaba, Baihetan, and Wudongde. All told, the combined hydropower output will be equal to two Three Gorges projects when completed in 2013.[57] In Sichuan, the Xiluodu hydropower plant, to be completed in 2017, is expected to generate up to two-thirds of the projected output of the Three Gorges Dam.[58]

---

55. Sanjeev Khagram, *Dams and Development: Transnational Struggles for Water and Power,* Ithaca, NY: Cornell University Press, 2004; and Conca, *Governing Water,* 167–68.

56. Andrew Mertha and William Lowry, "Unbuilt Dams: Seminal Events and Policy Change in China, Australia, and the United States," *Comparative Politics* 39, no. 1 (October 2006): 1–20.

57. "China to Build Four More Super-Large Power Plants in Upper Reaches of Yangtze River," *Xinhua,* September 30, 2005.

58. "China to Start Construction of Hydropower Plant," *Xinhua Financial Network News,* November 16, 2005. See www.threegorgesprobe.org/tgp/index.cfm?DSP=content&ContentID=12187, accessed July 27, 2007.

Hydropower is also an important source of clean, renewable energy completely under the control of Beijing. As such, it is firmly planted within one of the most politically charged and substantively important policy areas in China today:

> Sooner or later, virtually all of China's biggest problems come down to energy. For instance, making the poorer regions prosperous and relieving the rural-urban income gap—currently regarded as the country's biggest challenge—will naturally need more economic development, but maintaining the high rates of GDP growth will require the relieving of energy bottlenecks, which will itself rest on the constant construction of new power stations and the unrelenting search for new sources of oil and gas both in China and abroad. This itself creates problems relating to issues such as environmental protection and geopolitical security.[59]

China's bid through the China National Offshore Oil Corporation (CNOOC) to purchase Unocal and the ferocious debate it has engendered in the United States is just the most recent reminder that China's pattern of energy consumption has global implications. China is currently the second-largest consumer of energy, behind the United States. Given the current problems in the Middle East and China's independence from the U.S. campaign in Iraq and Washington's hostile posture toward Iran, China has leverage with some of the states in that region in a way that the United States does not. Moreover, China's appetite for energy is likely to increase dramatically in the years to come. Although China might have coal reserves that could sustain hundreds of years of future development,[60] the use of coal continues to wreak havoc on the environment, with the result that China boasts the dubious distinction of possessing sixteen of the world's twenty most polluted cities in terms of air quality, according to the World Bank.[61]

The health and other problems this has engendered, to say nothing of the threat to the international commitments that China has made with regard to environmental regulation, have led China's rulers to scramble for alternative sources of energy, including hydropower. The resulting mix of China's energy production and consumption of energy—which, the

59. "The Problems of Fueling Growth: A Review of China's Energy Industry in 2004," *Interfax,* January 14, 2005, http://interfax.com.

60. China has proven reserves of 126 billion short tons (80 years at current rates of consumption) and potential reserves of four trillion short tons (2,613 years at current rates of consumption). See www.cslforum.org/china.htm, accessed July 27, 2007.

61. China Environment Forum, "Smoggy Skies: Environmental Health and Air Pollution," Woodrow Wilson International Center for Scholars, February 13, 2007.

Table 1.3 China's coal production and consumption, 1993–2003, in short tons

Year	Production	Consumption
1993	1,304	1,276
1994	1,404	1,390
1995	1,537	1,495
1996	1,545	1,509
1997	1,507	1,450
1998	1,429	1,391
1999	1,365	1,343
2000	1,314	1,282
2001	1,459	1,357
2002	1,521	1,413
2003	1,635	1,531

Source: http://www.cslforum.org/china/htm.

proliferation of SUVs in the United States (and, more recently, in China) notwithstanding, remains a scarce and limited resource—will have global implications that are already beginning to be felt.

There is another important international dimension. Several of the rivers mentioned in this book originate in China but travel through Myanmar (Burma), Vietnam, Cambodia, and Thailand before discharging into the sea. Any alterations of the rivers in China (water level, flow rates, and so on) will have a substantial impact on navigation, irrigation, and the industrial uses of water from these rivers in the aforementioned countries. There are already rumblings of discontent from various groups in these countries. On the other hand, authorities like those in Yangon seem to be just as eager as those in Beijing to harness these rivers, and so they provide each other with some degree of political cover internationally. Finally, there have been extensive talks between Chinese agencies, such as the communications bureaucracy, with their Southeast Asian counterparts to manage the externalities of China's water policies.[62]

And, as noted above, this raises broader geopolitical issues. Insofar as domestic processes are able to halt the state's grand designs for harnessing China's hydropower capacity, China will more likely compete for a share of the diminishing pie of oil and natural gas, increasing the likelihood of exacerbating political tensions with the United States and the European Union and possibly even further warming its relations with Russia, which is now recasting itself as a global energy-producing superpower. The stakes are high indeed.

62. Interview 04KM04, August 20, 2004.

2 | Actors, Interests, and Issues at Stake

It is hard to generalise what the government thinks about us. The government is not a monolithic bloc in this regard. SEPA [the State Environmental Protection Administration] supports us and has called us their "natural ally." The MOWR [Ministry of Water Resources] probably likes us much less and the provincial government in Yunnan undoubtedly hates us.

—LIANG CONGJIE, founder of Friends of Nature, a Chinese NGO, quoted in Büsgen

In the past decade, China's energy demands have grown exponentially. As recently as the late 1990s, the United States used more energy by January 12 than China consumed in an entire year. Since then, demand for energy in China has skyrocketed, but supply has not kept apace. From 1980 to 2000, China's energy production increased at only half of its economic growth rate, but since 2001, energy growth has expanded to 1.5 times the rate of economic growth.[1] In 2003, there were serious energy shortages in two-thirds of China's thirty-one provinces as a result of a combination of poor planning, unexpectedly strong economic growth, and transportation bottlenecks.[2] The promise of oil and natural gas deposits that led to the rise of the petroleum clique in China in the 1970s and 1980s has not been borne out.[3] The legendary oil fields of Daqing have already seen their best days pass into the history books. The hundreds of mining disasters each year and the unparalleled pollution that hangs over every Chinese city have led China to scramble for alternatives. Yet China remains largely dependent on coal. In order to meet demand, local officials build on average a thousand-megawatt coal-fired power plant each week, adding the equivalent of Spain's entire electrical capacity every year—as a comparison, nuclear power plants

1. Douglas Ogden, "We Don't Need More Power," *Newsweek International*, February 6, 2006.
2. "China to Spend 50 Bln Yuan on 2 Stations for New Hydropower Project," *Three Gorges Probe*, March 22, 2004, http://www.probeinternational.org/catalog/three_gorges_probe.php.
3. Kenneth Lieberthal and Michel Oksenberg, *Policy Making in China: Leaders, Structures, and Processes*, Princeton, NJ: Princeton University Press, 1988.

typically produce between 500 and 2,000 megawatts when operating at peak capacity.[4] The sheer expense, risk, and time necessary to establish a larger nuclear energy infrastructure have led Beijing to proceed slowly with its "nuclear option." Nuclear power accounts for 1 to 2 percent of China's energy and is expected to rise to only 5 percent by 2030.[5]

Given these realities, an important energy source for China lies in its western hinterlands, in the form of hydropower. Although China leads the world in terms of its total theoretical hydropower resources at some 384 gigawatts,[6] by the end of 2004 China's hydropower capacity remained at the relatively modest level of 100 gigawatts, with ongoing projects to increase it by another 30 gigawatts.[7] Thus, China is currently only exploiting between 25 and 30 percent of its hydropower capacity, while the rugged province of Yunnan, which is 94 percent mountainous terrain, is only yielding 7 percent of its hydropower potential. In addition to the demand for energy in Yunnan and elsewhere in the country, hydropower exploitation in Yunnan also promises to alleviate the chronic poverty of the region, which encompasses some remote areas that reportedly did not make extensive utilization of the wheel (beyond religious and ritual use) until the 1930s. Finally, not only does China's hydropower potential render it completely free from external manipulation by its international neighbors and rivals, it is also a relatively clean technology and does not produce the CO_2 emissions associated with other forms of energy production.

But the policy area of hydropower is extraordinarily complex, pitting the growing number of actors opposing these projects against the technical uncertainties and potential negative externalities often downplayed by the hydropower proponents, some of which are simply in the business of collecting rents but many of whom genuinely believe that hydropower can help solve China's energy dilemma. Although the opinions are deeply held on all sides of the debate, dam proponents and opponents both have legitimate, even compelling, reasons for pursuing their interests. This conflict of ideas, already seemingly intractable, is made even more difficult by the politicization of otherwise objective points as well as by the corruption that undermines the arguments of hydropower proponents.

4. A watt is measured as one joule per second. A kilowatt is one thousand watts, a megawatt is one million watts, and a gigawatt is one billion watts. On the power plant, see Ogden, "We Don't Need More Power."

5. Interview 04BJ04, August 4, 2004.

6. Darrin L. Magee, *New Energy Geographies: Powershed Politics and Hydropower Decision Making in Southwestern China*, Ph.D. dissertation, University of Washington, 2006, 12.

7. Elaine Chan, "Three Gorges Just Act One in the Drive to Harness Nature: Planners Have Designs on Yunnan's World Heritage Area," *South China Morning Post*, November 7, 2005.

The Official State Frame: Economic Development

As an integral part of China's Tenth and Eleventh Five Year Plans, the "Develop the West" (*xibu da kaifa*) strategy provides the overall context of the political processes described herein. The western part of China accounts for 71.4 percent of China's overall landmass but holds only 28.6 percent of its population (364 million people in 2001). Even more striking, it only accounts for 17.7 percent of GDP. The overall plan focuses on infrastructure development, including railroads, gas pipelines, the west-east water transmission project, water exploitation, hydropower, and communications. This policy area is headed by the State Council Western Regional Development leadership small group (*Guowuyuan xibu diqu kaifa lingdao xiaozu*).[8]

The principal justification for the projects analyzed in this book is to develop hydropower capacity and the associated infrastructure as well as to bring local residents out of poverty by providing electricity and other positive spillovers from the associated regional development. Some dam proponents have gone so far as to say that forgoing such development risks depriving the local people of their very right to survival, citing the continuing degradation of the environment through deforestation and erosion brought about by slash-and-burn agriculture and similar techniques. Still others have placed the issue in political terms, such as "strengthening boundary security and stability" and fulfilling Jiang Zemin's "Three Represents" (*sange daibiao*) and "Three Stresses" (*san jiang*), particularly "stressing politics" (*jiang zhengzhi*). The newly paved roads along the Nu River in Yunnan make it easy to forget that less than five years ago, they were little more than dirt paths. The houses alongside these new roads still suggest almost unimaginable poverty. As recently as the mid- to late 1990s, many local residents had one set of clothes per family and had to take turns going out of the house.[9]

Beyond the rhetoric, however, the imperatives of economic development go far beyond bringing the people out of poverty. The economic contributions of these hydropower projects will be most directly felt in the increase of GDP and the expansion of tax bases for local governments.[10] Within the Nu River autonomous region alone, the Nu, Lancang, and Dulong rivers (and the 180 or so other first-grade branch rivers) have a water resource

---

8. State Planning Commission of the People's Republic of China and State Council Western Regional Development leadership small group office, "The Outline of [*sic*] Overall Plan for the Development of the Western Region during the Tenth Five-Year Plan Period."

9. Interview 05BS01, July 22, 2005.

10. Antoaneta Bezlova, "Let Public See Secret Mega-Dam Plans, Activists Say," *Inter Press Service/Global Information Network*, October 26, 2005.

capacity of 955.91 billion cubic meters, which could provide 20 percent of all the hydropower capacity in Yunnan province. The Nu River Project alone could provide twenty thousand megawatts of power (20 percent of China's current total) and 100 billion kilowatt-hours of electricity each year, the equivalent of 50 million tons of coal.[11]

There is some controversy concerning where the energy will go. The assumption is that much of it will remain in the area in order to help foster economic growth. In fact, much of it is to be sold to the coastal provinces such as Shanghai and Guangdong. This has also led to political squabbles over market inefficiencies. Guangdong province, for example, bristles at what its leaders see as a state-mandated reliance on southwestern energy. From 2001 to 2005, the central government banned hydropower projects in Guangdong in order to ensure demand for projects in the southwest, notably Sichuan and Yunnan. There seems to be less of a concern over pricing; rather, the issue of contention is over reliability in meeting market demand. Moreover, according to skeptics in Guangdong, seasonal fluctuations in the water levels in the southwest have not been considered.[12]

In addition, although dam proponents argue vaguely that there will be some sort of profit-sharing schemes whereby the resettled people (*yimin*) will share in the spoils, this remains largely a pipe dream. In the case of the Three Gorges Dam, the proposed profit-sharing (*fazhan fadian*) compensation is still being negotiated—thirteen years after construction began. Dam projects that are largely situated within a single province do not receive nearly the attention across the nation that the Three Gorges Dam did, and issues like the negotiation over profit-sharing schemes are taken even less seriously (see chapter 3).[13] Instead, there is usually a lump-sum compensation payout that covers some portion of the associated expenses and only the most equivocal promises of profit-sharing.

Resettlement

The official rhetoric about resettlement tends to be unabashedly upbeat. For example, the director of the Nujiang prefecture Party committee propaganda office, Duan Bin, argued that relocation provides opportunities to get rid of poverty. Once people have a nest egg, they can use the money to

11. Chan, "Three Gorges Just Act One."
12. Interview 05GZ01, November 4, 2005; Grainne Ryder, "Skyscraper Dams in Yunnan: China's New Electricity Regulator Should Step in," *Three Gorges Probe*, May 12, 2006.
13. Interview 05ZQ01, November 5, 2005.

develop the economy through tourism. Moreover, according to Duan, hydropower stations are beneficial to the environment because the relocation of the *yimin* will allow the forests to become virgin again.[14]

Yet resettlement remains a political lightning rod, and with good reason. Of the sixteen million people who have been resettled for hydropower projects in the past fifty years, ten million of them are still living in poverty.[15] In Yunnan alone, some put the total number of resettled people at half a million, with estimates that forty thousand people will have to be resettled every year for the next fifteen years.[16] Often, the vast majority of land slated to be submerged is high-yield farmland, while the eventual resettlement locations remain unspecified.

In general, the *yimin* in China have fared poorly. Zhou Tianyong, a professor at the Central Party School, undertook a survey of the *yimin* resettled as a result of the dams built along the Longyang Gorge in Qinghai province. Instead of reaping the benefits from the project, residents had an average net income of 1,772 RMB (US$220), one-fifth the national average, and 20 percent of these people had an annual income of 625 RMB or lower. Moreover, there is a strange disconnect between the local people and the dam projects built in their midst: "Although [the *yimin*] lived close to the dams, they did not have access to its water and relied on infrequent rains for drinking water. And power lines passed over their villages without sharing the electricity generated."[17]

As we will see, in the case of Pubugou in chapter 3, resettlement was the principal concern of the local peasants, while a postscript to the Zipingpu Dam near Dujiangyan, the focus of chapter 4, was the relocation of thousands of people, many of whom, six years later, have not been furnished with alternate accommodations and have been forced to move in with their extended families.[18] As mentioned above, high-yield farmland on the lower reaches of these rivers often stands in marked contrast to the poor-quality land along the upper reaches. Often there is no choice but to relocate the residents to another area altogether. This leads to other problems. First of all, initial development plans often account for only a portion of the actual

14. Cao Haidong, *Nujiang de minjian baowei zhan* ["The NGO Battle over Protection of the Nu River"] *Jingji*, May 2004.

15. Deng Jie, "Environmental Protection's New Power is Growing," *Southern Weekend*, December 27, 2005.

16. "500,000 People to be affected by Yunnan's Hydropower Development in the Coming 15 Years," *Kunming Evening Daily*, May 23, 2006.

17. Chris Buckley, "China Hydro-Dams Leave Locals Poorer," *Reuters*, February 15, 2006.

18. Pubugou Interviewee, October 31, 2005.

Figure 2.1 *"Yimin Banqian Gongzai Dangdai Li Zai Zisum"* (*Yimin* Resettlement for Success Today, Benefits for Posterity), Wangong village, Hanyuan county. Photograph by the author, August 2006.

number of people who eventually need to be relocated, and these development plans usually do not include any projections about the average farmland per person, the quality of farmland, land capacity, and other factors pertaining to peasants' livelihood. It is therefore impossible to compare these with the current living conditions of the *yimin*.

Second, the resettlement of the *yimin* also affects the social fabric of the locales in which they are resettled.[19] An official who assists in the relocation of the *yimin* from the Three Gorges underscored the complexity of this problem by outlining the myriad issues involved. These include the communication problems inherent in dealing with people who speak rudimentary Mandarin and a variety of mutually unintelligible local dialects. Furthermore, in the somewhat rare instances in which *yimin* are relocated to economically prosperous regions, they must make what is often a painful adjustment to a completely foreign economic environment. There are also the social tensions

19. Pubugou Interviewee, July 8, 2005.

that inevitably arise between the local residents and newcomers, especially if they are from a (non-Han) minority group.[20] Indeed, this official largely intimated that first-generation *yimin* would never fully blend into their new surroundings and placed his hopes on their children and grandchildren. Given that these are peasants, their experience outside their original locales is usually very limited, and their education levels are low. They often engage in extra-legal activities sponsored by gangs and secret societies because they do not understand or trust the official (or more widely accepted unofficial) channels, or else they join the ranks of the already-swelling "floating population" (*liudong renkou*). One Li minority villager who was resettled as a result of the hydropower construction on the Lancang River lamented that before resettlement it was possible to "leave the doors open at night" but that in his new home theft, prostitution, drug abuse, and all the other expected social ills have appeared.[21] The blame for these social problems is placed squarely at the feet of the *yimin*, further complicating successful assimilation into their new social environments. Not surprisingly, the "host" sites are often quite reluctant to accept *yimin*. Relocation assessments, such as they are, often do not take into account the preferences and concerns of the governments in the locations where these people are to be resettled.[22]

All of the foregoing is reinforced by a degree of separation anxiety that non-*yimin* can only imagine: being forced to leave towns and villages where their families have lived for hundreds or even thousands of years. This can have profoundly negative psychological effects. As one NGO government relations officer explained, this separation anxiety is also apparent in the strong superstitions embraced by many villagers in these underdeveloped areas. For example, when outsiders think about "sacred" mountains, they think that the local people go there to pray once a year, that it is a quaint idea. But for many of these villagers, the gods and ghosts that they believe inhabit these mountains and forests are a constant source of fear and dread. They believe that resettlement is much more than some process of exorcising the ghosts from one mountain or forest and "transferring" them to another. In fact, the villagers are in mortal terror of invoking the ire of

20. There are fifty-five recognized minority or ethnic groups in China, excluding the majority Han ethnic group. The non-Han minority groups make up about 123 million of China's overall population of 1.3 billion, or just under 10 percent. Over twenty of these ethnic groups are located in Yunnan province. See www.paulnoll.com/China/Minorities/China-Nationalities. html, accessed July 28, 2007.

21. *He gudiqu minzu wenhua de duoyang xinji qi zai shuidian kaifazhong de kunjing* ["River Valley Folk Culture Diversity and Its Difficulties with Water Resource Development"], *Kexue fazhan guan yu jianghe kaifa*, 38–49.

22. Ibid.; and Interview 05ZQ01, November 5, 2005.

these spirits by leaving the area. This adds another, intensely psychological dimension—one that defies valuation and compensation—to an already complex and complicated issue.[23]

Cultural Heritage Issues

Cultural heritage issues are another important unofficial issue frame employed by dam opponents. In addition to those discussed in the context of minorities in the next section, Han cultural heritage issues are also at play. One of the principal lightning rods of opposition in the case of the Three Gorges was the fact that the massive reservoir would inundate innumerable cultural relics, some dating back thousands of years. Although some of these were relocated, others have been submerged by the reservoir and are, for all intents and purposes, lost forever.

World Heritage designations figure prominently in this debate in that they give these threatened sites more visibility than would otherwise be the case. They also provide a degree of leverage for the opponents of hydropower plans. Although UNESCO plays a largely indirect role in the process, dam opponents have used UNESCO World Heritage designations as a way to embarrass Beijing by tacitly threatening to tell to the world that China does not care about its rich cultural heritage. As the Dujiangyan case suggests (chapter 4), cultural heritage issues are one notable area of interface among government agencies (the cultural relics bureaus), intellectuals, and activists, and they resonate strongly with the public.

However, cultural heritage issues are not a slam dunk. Besides the obvious fact that not all hydropower projects threaten world heritage sites, there are some serious limitations to the cultural heritage frame. First, the tactic seems to work better with man-made objects than with natural phenomena. Therefore, according to one peasant in Tiger Leaping Gorge, "nobody would ever consider knocking down the Great Wall, even as the government plans to fill Tiger Leaping Gorge with a reservoir."[24] Second, local governments have found ways to get around this issue, as in Yunnan province. The provincial leaders were able to negotiate that the Three Parallel Rivers World Heritage site only begins at a height of two thousand meters, therefore allowing the construction of any number of hydropower projects below that altitude, including those slated for the Nu, Jinsha, and Lancang rivers.[25]

23. Interview 05KM06, July 26, 2005.
24. Interview 05HTX03, July 17, 2005.
25. Interview 06BJ02, March 10, 2006.

Environmental Issues

The issue frame that was most compelling in the case of the Nu River has to do with the environment. The Three Parallel Rivers in Yunnan are responsible for some breathtaking scenery. The first bends of both the Nu and Jinsha rivers have 180-degree turns; in the case of the Jinsha, the bend is the topographical accident that ensures that the Yangtze, which the Jinsha becomes once it enters Sichuan, moves eastward instead of southward, providing the principal reason that the Chinese civilization, growing up around the unpredictable Yellow River, called "China's Sorrow," was able to thrive over the millennia. The entire area could be developed for tourism, as is the case with Lijiang and other culturally diverse towns scattered throughout the rugged terrain, including, but by no means limited to, Bingzhongluo, Zhongdian, and Gongshan. One of the principal issues for environmentalists is biodiversity, specifically the ecological environment of the areas under review. The Nu River Valley, for example, is home to seven thousand species of plants and eighty rare or endangered species of animals, and it has one of the highest concentrations of biodiversity in the world.[26]

However, one of the hallmarks of the Nu River opposition movement has been its shift from simply focusing on the environment to framing environmental issues in terms of the local population and to good governance issues, what activists refer to as a "shift from biodiversity to people." Green Watershed, an NGO in Yunnan, has focused on good governance as much as it has on watershed protection. This shift to the livelihood of the people has expanded to include biodiversity and socio-cultural diversity, particularly in Yunnan. Given Yunnan's rich minority culture—some four-fifths of China's minority groups can be found in the province—opponents have bundled minority issues together with environmental issues. Although they have rarely brought up minority cultural issues in the absence of environmental concerns, they have often fused the two. It seems that the latter, taken alone, does not have the necessary traction. When proposed in conjunction with environmental issues, however, it does seem to add considerable weight.

The three cases discussed in this book share the characteristic of seismic instability.[27] Lijiang, which is near the Jinsha River, where a network of hydropower dams is being proposed, was partially destroyed in 1996 by an earthquake measuring 7.0 on the Richter scale.[28] Along the Nu, there are

26. Jim Yardley, "Dam Building Threatens China's Grand Canyon," *New York Times*, March 10, 2004.

27. Interview 05CDA, July 8, 2005.

28. See www.worldbank.org/yunnan/overview.htm, accessed July 29, 2007.

frequent landslides, mudslides, and falling boulders. Throughout much of the year, the roads that hug the Nu River are impassable. In fact, the Nu runs along a major fault line, which forms the two major earthquake zones of Yunnan province, the Yunnan Southwest and Tengchong zones. Seismic activity grows gradually stronger from the middle reaches of the Nu river to its lower reaches. Critics have argued that at no time during the planning process were these concerns taken into account. The same geological concerns are present in the cases of Dujiangyan and Pubugou (the latter are even suggested on the Hanyuan county website).[29]

There is also the problem of silting. The principal component that makes up the main body of silt in the Nu River is the large rock debris from the mud and rock flows and landslides.[30] In the absence of dams, this debris is gradually pushed down to the lower reaches, which results in the river bed being balanced. But once these dams are constructed, much of this debris will be caught and accumulate in the reservoir, thus seriously affecting the duration of use of the dams and their overall efficiency.

The Context: A Glimpse into the Complexity of Local Interests

In moving the discussion to state institutions, it is important to mention at the outset that China is by no means a unitary state. Thus, it is important to move away from generalizations such as "the state wants" or "the state believes" because "the state" itself comprises an extremely complex set of preferences, goals, and authority relations. Aggregating these to something we refer to as "the state" or "China" causes us to lose valuable information and fundamentally misunderstand the structure and process of Chinese politics. Rather, it is important to disaggregate the various component parts, keeping the epigraph of this chapter firmly in mind. Another caveat is in order. The exercise of dividing these actors into two camps, the state and society, may assist with conceptual clarity, but it does a disservice to the complexity of interests. Before discussing the proponent/opponent divide below, I offer a glimpse into some of the layers of incentives and interests that make this issue so difficult to simplify.

29. See www.yahy.cn/Index.html, accessed on July 29, 2007.
30. On both of my trips to the Nu River, in July 2005 and March 2006, I was delayed for several hours by mudslides. Along the lower trail of Tiger Leaping Gorge, many areas of the paved roadway have what look like big "bites" taken out of them as a result of falling boulders from above. This does in fact lead to a few dozen deaths in the area every year.

When asked what they think about the Nu River Project (NRP), local people's reactions have been decidedly mixed. Many people simply do not know about it. This varies depending on how near or how far one lives from the banks of the river: the closer one is, the better informed one is, but the effect is slight; in remote areas, 75 percent of the people know nothing about it. Those familiar with the NRP—government officials, CCP members, teachers, postal workers—frequently parrot the official line of economic development. But when one asks more probing questions, they get confused and become unsure of exactly what to say. Another category of people, including dam workers and business people, sees opportunities with the NRP. There are some people, however, who are skeptical, such as the landowners. But for the most part, it is difficult to get people to say that they are opposed. People usually hear about such projects through local newspapers and word of mouth. They often repeat what they read and hear—that it is good for economic development.

However, it would be inaccurate to state unequivocally that all of the locals are simply being duped. The local people very much depend on their village and township governments, especially if they are labeled "national poverty areas," like those in Fugong county. Local governments give the people handouts and other benefits such as "effectively free" education through high school (depending on scores), subsidies for housing, irrigation, and drinking water. Moreover, in cases like the 2004 floods in Simao prefecture, the local government worked extremely aggressively and effectively to help the affected citizens. The people remember this, and it seems that their hearts and minds were won over for some time. One often hears that the local leaders of the Kuomintang (KMT) government in the area in the 1930s and 1940s were particularly unconcerned with how they were perceived by residents—people still recall that KMT officials insisted on being carried on litters wherever they went. The "peaceful liberation" of the area from abusive warlords and corrupt KMT officials by the CCP is by no means an empty euphemism. The latter has maintained generally good relations with the local people with relatively little conflict, although the area did not escape the devastation of the Cultural Revolution.

In fact, there seems to be—in Fugong at least—"a lot of appreciation and trust." So if the NRP provides more revenue to the local government that will then be distributed locally, the locals will support it. This, of course, depends on maintaining "manageable" levels of corruption. Local officials certainly skim off the top, but as long as such corruption stays within a manageable level, it is tolerated. The conventional wisdom among villagers is that local officials have a terrible job with very little compensation, so the cadres deserve a little more. National-level funding is channeled through the province (Yunnan), the prefecture (Nujiang, or "Nu River"), the county (Fugong), and to the villages and townships, and most of it "seems to be reaching its destination."[31]

In other areas, such as the Tiger Leaping Gorge along the Jinsha River, where a proposed 912-foot dam would create a reservoir for 125 miles to assist in water storage and regulate flow to the Three Gorges Dam. Downstream, where the Jinsha turns into the Yangtze, corruption is rampant, and the local people are notably outspoken about it.[32] In Haiza village, a family told me that they and all their neighbors strongly opposed the proposed hydropower project. They said that they would be relocated and held out little hope of adequate compensation. The local government already takes one third of their income—as it does for everybody in the area—and they fully expect that the local government will do the same when it comes to any kind of compensatory arrangement they receive for relocation.[33] A peasant entrepreneur at a higher altitude echoed these sentiments. He immediately volunteered that the local leaders were corrupt, saying that he had attended meetings to "discuss" the dam project where he and the other villagers were presented with what was effectively a fait accompli.[34] Some people, like the farmers around the town of Shigu adjacent to the first bend of the Jinsha River, have refused engineers access to the area necessary for initial surveying work, while others are fatalistic or even welcome such a development: "The dam will bring progress for us. And tourists will still come, if not to look at the gorge, then to look at the dam just like they visit the Three Gorges Dam."[35] In short, there is tremendous variation within the localities affected by these developments.[36]

Actors: The Pro-Hydropower Forces

The case study chapters to follow provide "critical cases" in which policy change as a result of bottom-up pressure is least likely to occur. This is because of the importance of the issue and the immense political power of

31. Interview 06FG01, March 13, 2006.
32. Jane Macartney, "Greed for Energy Threatens to Dam Legendary Gorge," *The Times* (London), May 9, 2006.
33. Interview 05HTX01, July 16, 2005.
34. Interview 05HTX03, July 16, 2005.
35. Macartney, "Greed for Energy Threatens."
36. I have focused on local people in this section; to avoid repetition, I discuss the ambivalence of governmental units in the sections that follow.

the more ardent supporters of this policy. The political stakes are arguably as high as those at play during the Three Gorges debate during the 1980s.[37] The dominant official frame for all three cases is the policy "line" of "Develop the West," a key goal of which is to alleviate the economic imbalance between China's poor interior and rich coastal areas. The policy predates the rise of Party General Secretary Hu Jintao and Premier Wen Jiabao, but it has become linked with their populist calls for overturning the extreme inequalities associated with economic development over the past twenty years.[38] In the words of one official,

> Even if foreign investment does not meet its projected goals, the "Develop the West" strategy will continue. This has to do with economics, but it is not simply an economic problem. In reality, it is a social problem and that makes it a political problem.[39]

Hydropower stations provide electricity that can be used locally or sold to China's richer provinces, with the profits reinvested into the original localities. Hydropower stations require an infrastructure (roads, bridges, and so on) that assists in the economic development of these poorer regions. Shipping within the interior, made possible by the damming of these rivers and the locks provided by these dams, increases the import and export of goods and commodities to and from these locales. Finally, hydropower stations, more specifically the reservoirs that are part of them, help mitigate against flooding and bring about a greater degree of regularity and predictability to farmers located downstream from these rivers. Thus, China's "western strategy" has become an important pillar of the modernization agenda and a standard against which to measure the success and, by extension, the legitimacy of China's current ruling elite.

The Water Resources Bureaucracy

The Ministry of Water Resources (*shuili bu*, MWR) was established on April 9, 1988, taking over from its predecessor, the Ministry of Water

---

37. The Three Gorges Dam Project was designed to provide flood control to alleviate the endemic flooding of the Lower Yangtze Basin, to provide irrigation for upriver areas of the Yangtze, and to allow shipping of up to ten thousand tons to the interior entrepôt of Chongqing. Each of these goals contributes to the development of China's poor interior provinces of the Southwest (Sichuan, Guizhou, Yunnan, and, since 1997, Chongqing). The decision to move forward was championed by Premier Li Peng, who was trained as a hydropower engineer in the 1950s and who had previously been the minister of water resources.

38. China's Gini coefficient has grown to 0.465 in 2004, indicating a level of inequality on par with that of the United States. See en.ce.cn/Insight/200509/13/t20050913_4669147.shtml.

39. Interview 05GY01, July 9, 2005.

Resources and Electric Power, formed in February 1958. The MWR was upgraded in rank in March 1993. The individual bureau-level officials of the MWR system (*xitong*) appear to be equal to other, related bureaus such as the infrastructure and land management bureaus, but these MWR bureaus are in fact regarded as *zucheng* (integral component units); even though the MWR offices cannot issue binding orders to these other units, they do remain a half-step above them in the byzantine network of Chinese bureaucracies. At the national level, the MWR has seven water commissions (WC) tasked with managing the river projects along China's major rivers that cross provincial boundaries and which have corresponding offices at the provincial level: Huai River WC (HRWC), Changjiang (Yangtze River) WC (CJWC), Zhujiang (Pearl River) WC (PRWC), Huanghe (Yellow River) WC (YRWC), the Songliao WC (SLWC), the Heilongjiang (Black Dragon River) WC (BDRWC), and the Taihu River Basin Management Department. The provincial counterparts of the CJWC, for example, are the water resources (WR) bureaus of Sichuan province and Chongqing, a provincial-level municipality. Each of these bureaus handles its part of the Yangtze River, with overall coordination by the CJWC, which has non-binding professional, consultative relations (*yewu guanxi*) with each of these. If a river is largely contained within a province or subprovincial locality, it is managed by the flood control department (*fanghong chu/fangxun ban*).[40] The relationship between the MWR and these watershed commissions is complex and somewhat counterintuitive:

> Like the other six basin commissions, the CJWC is theoretically subordinate to the MWR, yet in actuality it has been delegated authority by the State Council *to approve or reject projects based on whether or not they meet the requirements of the comprehensive plan for a particular river basin.* For instance, projects that abstract more than 100,000 m³ of water per day on the upper Yangtze (Jinsha) must all be approved by the CJWC, not by the government of the province...nor by the authorities in the Ministry of Water Resources or the State Council.[41]

Not surprisingly, this has created a fair amount of confusion in terms of authority relations.

Below the national level, with some variation, the water resources bureaus largely conform to the following contours. They are, first of all, "first-tier bureaus" (*yi ji ju*). They have different departments that vary across counties, depending on the physical and topographical characteristics

---

40. Interview 05KM02B, July 20, 2005.
41. Magee, *New Energy Geographies,* 241, emphasis in original.

of the area. There are no strict rules in how they set up departments, and they have a fair degree of freedom in this regard. This bureaucracy is decentralized (*kuaishang lingdao*), as it receives its budgetary and personnel allocations (*bianzhi*) from the government at the same corresponding administrative level, not from its functional administrative superior. It has units all the way down to the township and village level—at this level, they take the form of WC "stations."

In terms of the application process, provincial WR bureau approval is necessary if the volume of the dam is ten million cubic meters or if the dam height is more than fifty meters. If the volume is one to ten million cubic meters or if the height of the dam is between thirty and fifty meters, the prefecture water resources bureau must give its approval. If the volume is one hundred thousand to one million cubic meters or if the height is between fifty and thirty meters, the county-level water resources bureau must give approval. If the dam project extends into another county or prefecture, then the water resources bureau at the next-highest administrative level (prefecture for a county, province for a prefecture) must sign off on the project.

There are three systems for approval. They are decided by the provincial water resources bureau, which then establishes the standard used. The first is "examination and approval" (*shenpizhi*) and is used if the project is "very big or important" at the provincial or even the national level—the *shenpizhi* standard is very strict and formal. The second is "authorization" (*hezhunzhi*) and is usually reserved for projects that are smaller. The procedure is somewhat looser, but it is still necessary to receive official and documented approval from the water resources bureau—it is formal but less strict. The third is "recordal" (*bei'anzhi*), in which it is not necessary to obtain prior approval, only to make a record of the project; this is usually the case with much smaller ventures. Projects undertaken by the national-level Huadian and Huaneng corporations, discussed below, are usually so large in scale that they do not get approved at the provincial level, but rather from Beijing.[42]

Although the water resources bureaucracy is often singled out as the most avidly pro-hydropower one, it is actually outflanked in this regard by the somewhat less "narrowly professional" and more "ideological" and "macro-oriented" National Development and Reform Commission. The former handles dam and hydropower issues, while the latter incorporates them into the larger scheme of economic development in China writ large.[43]

42. Interview 05GY04, July 11, 2005.
43. Interview 05KM02B, July 20, 2005.

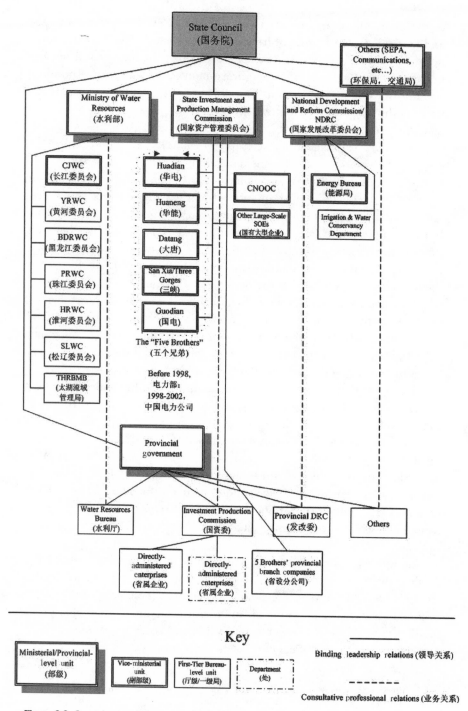

Figure 2.2 Organizational structure of the three principal hydropower *xitong*.

The "Economic Helmsman": The National
Development and Reform Commission

The National Development and Reform Commission (NDRC, often referred to in its abbreviated form, *fagaiwei*) is a tremendously important
player in this policy area. It draws its immense power in part from its institutional legacy—it was formerly the State Planning Commission (SPC),
charged with the overall economic and infrastructure development of the
country. In 1998, the State Planning Commission was renamed the State
Development Planning Commission. In 2003, it was merged with the State
Council Office for Restructuring the Economy and parts of the State Economic and Trade Commission and recast as the NDRC.

Even though some anticipated that the NDRC would be weakened in the
move toward a market economy, it has in fact retained its authoritative position within the Chinese government. Indeed, the NDRC, in the parlance
of national officials, is often referred to as the "small State Council" (*xiao
guowuyuan*) because of its tremendous power and administrative range. Traditionally, commissions unambiguously outranked ministries, and although
there have been recent trends to bring commissions closer to the administrative rank of ministries, because of their coordinating functions—they
coordinate ministries within their functional area of operations—commissions
retain the power and privileges of a half-step increase in bureaucratic rank
over ministries and provincial governments. However, if all commissions
enjoy equal rank with one another, to paraphrase Orwell, some are more
equal than others. And the NDRC enjoys a degree of power that many other
commissions can only dream of.

Today's NDRC is composed of two of the most important traditional
commissions in China, with all of the institutional baggage such a com-
posite unit necessarily possesses. Indeed, the NDRC appears to encompass
two related yet seemingly contradictory dynamics. On the one hand, the
NDRC has decentralized—or in the view of one observer, "disintegrated"—
organizationally while maintaining or even enhancing its own political
power. In the past (except during the Cultural Revolution), the NDRC (or
its forerunner, the SPC) would hold fairly regular annual meetings to discuss big-ticket investment and development projects. Throughout the Reform Era, however, there has been a trend away from such formal meetings;
rather, the tendency has been in the direction of more informal meetings
among ranking cadres and the issuance of approvals for projects in which
they have a personal stake, but with the imprimatur of the NDRC.[44]

---

44. Ibid.

In addition, there has also been a tendency to move an increasing number of nonexperts, including ever-expanding hordes of unqualified military and CCP cadres, into key NDRC positions.[45] Yet given their political power—which these trends only enhance—they cannot be fired for corruption, let alone incompetence or negligence. This makes it less likely that the NDRC will judge a given project objectively, on its merits.

The administrative rank of these offices underscores the tremendous political power of the NDRC and its local counterparts and their ability to shape the broad contours of development policy. The director (*zhuren*) of the NDRC is extremely powerful, even more so than is suggested by his formal title, which has the administrative rank just above minister or ministry, *bu ji*. However, the members (*yuan*) on the commission (*weiyuanhui*) are either ministers or those with the lower director- (bureau-) level rank (*ju ji*).[46] In order to coordinate these other committee members, the director of the NDRC must have a higher rank. Frequently, the next step on the career ladder for a director of the provincial DRC is that of vice governor or provincial vice party secretary or a position of like administrative rank.

But there is a more substantive reason for the NDRC's enormous power—its immensely broad functional jurisdiction and policy area. Indeed, one source argued that—at the local level, at least—the NDRC's political maneuvering can easily compete with that of the local government, the jurisdiction of which the NDRC is formally under, since the DRC system is based on decentralized leadership relations. That is because the NDRC encompasses those administrative units that are at the very forefront of economic reform in China, as they are not dragged down by the many other units that serve more of a social welfare function, which represent drain of budgetary and other resources that generalist government units, by definition, have to manage. The units that are under the coordination of the NDRC include water resources, hydropower, and communication/infrastructure agencies, among many others.

Any local project that is in any way related to power stations must go through the DRC Energy Bureau (*nengyuan ju*) at that same administrative level. This is a second-tier (*futing ji*) unit within the DRC; it ranks a half-step above the Irrigation and Water Conservancy Department (*nongtian shuili chu*), which is a third-tier (*xian ji*) unit within the DRC at the national and

---

45. Ibid. This is not unlike the case of the Administration for Industry and Commerce bureaucracy (*gongshang xingzheng guanli xitong*), which was forced to absorb a great many of the soldiers demobilized in the early 1980s.

46. RDCs extend all the way to the county level, where they are referred to as *fazhan ju* ("development bureaus").

provincial levels; this means that hydropower concerns effectively trump those of irrigation and water conservancy to some degree.[47]

Another thing that distinguishes the NDRC is that water resources units are often (though not always) more objective than the NDRC. This is because the water resources units have several water-related responsibilities that extend into the medium to long term. The NDRC's mandate is: development *now*. Thus cadres in the NDRC are promoted based on these short-term projects, while those in other units have incentives that are far less immediate in terms of their own prospects of promotion.[48]

The Five Brothers (and Their Smaller Cousins): The Hydropower Companies

Obviously, the hydropower companies are also fiercely in favor of energy development. Between 2000 and 2002, there was, in the words of one expert, a "hydropower rush" (*xi he quan shui*, literally "to occupy the river") during which economic and political incentives for these companies, their political sponsors, and local governments were sharply skewed in favor of construction at all costs.[49] The evolution of these companies also betrays the degree to which they encompass political interests.

In 1996, the Ministry of Electric Power was "put on notice"—it was finally abolished two years later—and in its stead arose the State Power Company of China (SPCC). Former cadres of the Ministry of Water Resources were now working for the SPCC. The individual entities within SPCC (*Guodian*) sought what they euphemistically referred to as "increased competition and fewer monopoly constraints"—in other words, the ability to invest as they wanted. But it was considered too unwieldy to be managed effectively and did indeed get embroiled in a high-profile scandal resulting in the flight of its director, Gao Yan, to Canada.[50]

So in 2002, SPCC's assets were redistributed among five electrical power companies, the "Five Brothers" (*wuge xiongdi*): Huadian, which owns a majority stake in the Yunnan Nu River Hydropower Development Company, the owner of rights to build on Yunnan's Nu River; Huaneng Group, the largest of the five, which owns a majority stake in Hydrolancang, which in turn holds development rights in the Lancang River in Yunnan; Sanxia, which holds responsibility and rights to building dams along the upper

---

47. Interview 07KM01, March 12, 2007.
48. Interview 05KM02B, July 20, 2005.
49. Interview 05KM03A, April 28, 2005.
50. Erik Eckholm, "Chinese Power Company Chief Flees the Country and Scrutiny," *New York Times*, October 20, 2002.

reaches of the Yangtze—the Jinsha River; Guodian; and Datang.[51] The directors of these corporations have the administrative rank of a vice minister or provincial vice governor. They are embedded within the State Investment and Production Management Commission, which manages other large state-owned enterprises in other economic sectors.

These corporations may formally be ranked at the national level as vice ministries, but the individuals who are the actual directors of these corporations personally have ministerial rank. Perhaps the most notorious instance of political power at the helm of these units is the director of Huaneng, who is none other than Li Xiaopeng, the son of Li Peng, former premier and chairman of the Standing Committee of the National People's Congress and, not incidentally, Three Gorges patron. Each of these directors personally has such a background, making these corporations more powerful than their formal ranking would suggest. In the old days, NDRC approval was necessary for releasing the money for these state-owned enterprises that handled hydropower projects, and in theory, the same holds true today (so Huaneng would have to get approval from the CJWC, which, in turn, would secure approval from the NDRC). But in practice, because these companies have so much money (as do many of the government units that are involved in these projects), they often go ahead without prior NDRC approval.[52]

After these five companies were established, each wanted to build up its business as rapidly as possible. Each company's strategy was to go to the part of the country that was most underdeveloped, the southwest. Moreover, because these areas were so poor, the local governments were very eager for any kind of big investment on the theory that big projects lead to even bigger investment.[53] Furthermore, local officials had negative incentives to pursue such a strategy; that is, if they didn't bring in such investment, they would be sacked.[54] The companies have representative offices at the provincial level all over China as well as at the sites of the projects in which they are involved. In Yunnan, for example, Huadian manages projects along the Nu River, and Huaneng handles those along the Lancang River (through the Huaneng Lancang River Corporation, which is itself split up into smaller companies). The Yunnan provincial government has a 30 percent stock ownership in

51. Magee, *New Energy Geographies,* esp. ch. 7; and Ryder, "Skyscraper Dams in Yunnan."
52. Interview 05KM02B, July 20, 2005.
53. David Lampton, "A Plum for a Peach: Bargaining, Interest, and Bureaucratic Politics in China," in *Bureaucracy, Politics, and Decision Making in Post-Mao China,* ed. Lieberthal and Lampton, 33–58.
54. See Kevin J. O'Brien and Lianjiang Li, "Selective Policy Implementation in Rural China," *Comparative Politics* 31, no. 2 (January 1999): 167–86.

Yunnan Huaneng operations, with national Huaneng owning the other 70 percent.[55] This allows the provincial government to have *some* influence but not too much. Rather, the national-level Huaneng Group has centralized leadership relations (*tiaoshang lingdao* or *chuizhi guanli*) with Huaneng Lancang River Corporation at the provincial level in Yunnan.[56] The minister of water resources just before the latter was transformed into the SPCC was Gao Yan, who also happened to be the former Party secretary of Yunnan Province. It is reasonable to believe that he pushed for a large presence of what eventually became local Yunnan Huaneng and Huadian operations.

In addition, since 1992 there have been many companies, such as Yunnan's Hongta Group and development foundations, that have cooperated with the Yunnan Hydropower Company Group (YHCG), which was converted from a government unit to a state-owned enterprise after 1997. These companies were invited to work on the other Lancang dam projects—Dacaoshan (1997–2002) and the Xiaowan (as yet unbuilt)—and established the Lancang River Hydropower Company. At this point, there were no longer any financial constraints.

This "hydropower rush" has attracted a motley crew of investors of all shapes and sizes who have underwritten large-scale dam projects as well as those along smaller, tributary rivers. The investment is of two types. The first type of investor purchases the right to build the hydropower stations. The second purchases the right to sell the electricity generated by the hydropower stations. In the past, projects like this were exclusively in the investment domain of large companies that reflected government offices' interests. Now, these large companies, which mostly invest in the main parts of, say, the Nu River Project, must receive national-level approval and their investment schemes are handled by the national government offices. The large companies have been joined by smaller private investors. These smaller companies, who generally invest in projects along the smaller rivers, negotiate the terms with the local governments' power bureaus (*"sheng ji xia de [zhou xian] zhengfu dianli bumen"*).[57] Thus, the diversity of the type of hydropower investors is equally matched by the many channels of government-business relations in which they engage.

That these companies had no experience with dam building was largely irrelevant. But this raises the question of why these companies—particularly the Hongta Group, which had made its fortune from Yunnan

55. Interview 05KM02A, April 27 and 28, 2005.

56. On *chuizhi guanli*, see Andrew C. Mertha, "China's 'Soft' Centralization: Shifting *Tiao/Kuai* Authority Relations," *China Quarterly* 184 (December 2005): 791–810.

57. Interview 05KM01, April 26, 2004.

tobacco—moved so rapidly into hydropower investment. Tobacco taxes have traditionally been collected as a national tax (*guoshui*), but in the case of Yunnan, they were allowed to be kept in the region in the form of a local tax (*dishui*). During the 1990s, the national authorities became aware of the sheer volume of revenue to be drawn from Yunnan tobacco, and these taxes were converted into *guoshui*, which meant the Yunnan government saw a significant portion of its tax base dry up. There followed a scramble for new sources of revenue—and hydropower was an obvious candidate.

The Manwan project was the biggest asset for the YHGC. When the Manwan dam was built, Yunnan owned 30 percent of the assets and the national government owned the balance. Once the Five Brothers were created, Manwan went to Huaneng, smaller dam projects went to Huadian, and the electricity transmission lines were given to (or stayed in) the YHGC. They kept the prices artificially low to prevent investors from seeking out smaller electricity transmission lines that were not part of the southwestern energy grid. By keeping the prices low, it was also possible to consolidate the grid. At this time, the regulatory framework was still very weak, while the hydropower sector was wide open, causing a scramble for investment.[58] It still resembles the Wild West in some ways, due in no small part to the influence of maverick local governments in an era of decentralization.

It should be noted that local hydropower companies—that is, those independent of the Five Brothers—are among those who oppose large-scale hydropower construction because it interferes with their own more modest project goals. These smaller operations tend to fall beneath the national radar and are harder to regulate, adding an additional wrinkle of complexity to the issue. For example, the Zhaoting municipality (Yunnan) DRC plans to invest up to 14 billion RMB to construct a set of hydropower stations with a combined output of 2 million kilowatts.[59]

Local Governments and Others

Most local governments tend to side with the hydropower projects, usually due to economic considerations as well as political prestige for the localities and potential career advancement for individual officials. Generally speaking, the more remote the region in question, the more able the local government is to control information. And they exploit this advantage masterfully, sometimes through coercion and force.

---

58. Interview 05KM03A, April 28, 2005.
59. "Yunnan to Invest CNY 32.5bn in Power Construction," *China Construction Project Net*, July 5, 2005.

Figure 2.3 Red flag over the Nu River: Guodian banner at the Yabiluo Dam site. Photograph by the author, Yabi-luo, March 2006.

Another demographic that is fiercely pro-hydropower is the sector of more than a million people whose livelihood is directly related to dam construction. This group actively lobbies for new projects. Indeed, a large professional class of hydroelectric engineers help lubricate the state's drive for increased hydropower: even as a project is being undertaken, these engineers are already active in promoting future projects that will keep them gainfully employed. With an eye on the completion of the Three Gorges Dam Project in 2009, they are already scrambling for work. As the chief engineer of the Zipingpu project told me when asked how many dams he had already built, "twenty or thirty, one dam at a time...." It is impossible to put this group of people out of work, just like that. In fact, this demographic is an important but often overlooked part of the supply side of the issue.[60] As former engineers, many of China's leaders share a

60. Interview 04BJ05, August 4, 2004.

professional affinity with this group and often provide a sympathetic ear to their demands.[61]

Actors: The Anti-Hydropower Forces

Elements within Local Governments

Although the pro-hydropower forces possess formidable power, their victory within the policy process is no longer assured. Although this is somewhat counterintuitive and contrary to much of the conventional wisdom, not all local governments are pro-hydropower—or, at least, pro-*big* hydropower.

SEPA and Local EPAs

The State Environmental Protection Administration (SEPA), is a relatively new bureaucracy that, like many such bureaucracies, is lean, and, in a word, aggressive. It is also somewhat short of revenue streams, so it is always looking for programs and possibilities to augment its power (or to capture more operating funds). Part of SEPA's power resides in the individuals that run it (which is difficult to replicate locally). In particular, Director Zhou Shengxian and Vice Director Pan Yue have used their bully pulpits to become vocal proponents of SEPA's institutional mandate and organizational goals. Pan in particular is perceived to be a rising star, because he has been elevated to this level while only in his mid-forties. Pan has a background in journalism, which he exploits masterfully, and he also used to work in the Economic Restructuring Office. Pan's father is an engineer general in the People's Liberation Army, and his father-in-law is Liu Huaqing, so he has strong personal connections as well.[62]

Pan has applied an entrepreneurial approach to compensating for some of SEPA's institutional weaknesses. For example, he has helped SEPA organize environment-based student groups at more than 170 universities.[63] SEPA may not necessarily be "one of the weakest ministries in the central government," but it is often outflanked by many stronger opponents.[64] Below the provincial level, the environmental protection bureaus (EPBs) do become quite weak.[65] In fact, EPBs tend to get even weaker as one goes

61. Interview 04BJ04, August 4, 2004.

62. Interview 05BJ01, July 4, 2005; and Interview 05BJ04, July 5, 2005.

63. Jim Yardley, "Bad Air and Water, and a Bully Pulpit in China," *New York Times*, September 25, 2004.

64. Yardley, "Vast Dam Proposal Is a Test for China," *New York Times*, December 23, 2005.

65. Interview 05BJ01, July 4, 2005.

Figure 2.4 Dam construction at Dimaluo, Gongshan county. Photograph by the author, March 2006.

down the administrative hierarchy—they retain the title of *huanbao ju* at the prefecture and county levels, but at the township and village levels, they have the somewhat weaker designation *huanbao yuan*. The entire hierarchy is decentralized.[66]

At the local level, in Yunnan, the EPB was a second-tier bureau until around 1993, when it was promoted to first-tier bureau status at the same time that SEPA in Beijing was upgraded to the ministry level as a *zong ju* or *bu ji bumen*.[67] The timing was not uniform throughout China, but by 2007, all provincial-level EPBs had become first-tier bureaus. However, in the arcane world of Chinese bureaucratic politics, as mentioned earlier, some bureaus are more equal than others. For example, the provincial-level EPBs

66. Interview 05GY05, July 11, 2005.
67. On the administrative rank of bureaus within the provincial government, see Mertha, *The Politics of Piracy,* esp. ch. 3.

are first-tier bureaus, as are the water resources bureaus. But provincial-level administrative organs (*sheng zhengfu gongzuo bumen*) are further divided into several subtypes: administrative offices (*bangongshi*), directly administered special offices (*zhishu teshe jigou*), directly administered offices (*zhishu jigou*), longstanding offices for coordinating and discussing official business (*yishi xietiao jigou de changshi banshi jigou*), and managing offices/units (*bumen guanli jigou*). Finally, there are the somewhat more powerful organizational units (*zuzhi bumen*). The EPBs are directly administered offices, while the water resources bureaus are organizational units; although the two have the same formal administrative rank, in terms of the hierarchy of responsibilities the water resources bureaus, as *zuzhi bumen*, are regarded as somewhat more influential.[68]

At the prefecture level, EPBs were separated out of the construction bureaus and transformed into "independent" (*duli*) units in 1997. Even when these two units were merged, it was not a relationship described as "one office, two separate signboards," but one office with "EPB" always following "construction," indicating the secondary importance of environmental protection. Although there are some efforts underway to centralize the EPB bureaucracy, it remains decentralized.[69] That is, a provincial-level EPB receives its personnel and budgetary resources (*bianzhi*) from the provincial-level government, which has sole authority relations over the provincial-level EPB. And this continues down the administrative hierarchy.[70]

The *Wennwu* (Cultural Relics) *Xitong*

Armed with the Cultural Relics Law, which was adopted on October 28, 2003, the cultural relics bureaucracy ensures that roads, dams, reservoirs, and other new infrastructure projects do not adversely affect China's cultural and historical relics. Given this mandate, the national-level Cultural Relics Bureau is, on balance, opposed to such hydropower projects. The cultural relics provisions are the oldest of the original Culture Law, dating back to 1977 at the time immediately following the fall of the Gang of Four and enacted as a reaction to the devastation of the Cultural Revolution.[71]

---

68. Interview 05GY04, July 9, 2005; and Interview 05GY05, July 11, 2005.
69. See Mertha, "China's 'Soft' Centralization"; and Martin Dimitrov, "The Politics of Selective Centralization and Decentralization in China: Evidence from Intellectual Property Rights, Environmental Protection, and Securities Regulation," paper presented at the Association for Asian Studies Annual Conference, April 2, 2005.
70. On *bianzhi*, see Kjeld Erik Brødsgaard, "Institutional Reform and the *Bianzhi* System in China," *China Quarterly* 170 (June 2002): 361–86; and Mertha, "China's 'Soft' Centralization."
71. Interview 05KM01, April 26, 2005.

At the national level, the Cultural Relics Management Bureau (*guojia wenwu guanli ju*) is a vice-ministerial ranking unit within the Ministry of Culture. This means that the director of the former is also a vice minister of the latter. There are some provinces so rich in cultural heritage sites that they are referred to as *wenwu da sheng* ("culturally-rich provinces"—these include Shanxi, Shaanxi, Hebei, Beijing, and Tibet). In these provinces, the Cultural relics bureaus (CRB) and the culture bureaus (CB) are both at the same first-tier rank; in other words, they are close but independent in terms of personnel/budgetary allocations and office resource outlays (*rencaiwu*).[72] In other provinces, such as Sichuan, Fujian, Zhejiang, Jiangsu, and Yunnan, the CRBs are second-tier units. The relationship between the CB and the CRB is one of "two separate units with two separate signboards" (*liangge jigou liangkuai paizi*). One source added that at the front gate, the relative power relations are symbolized thus: "you will notice that the *wenwu paizi* [cultural relics signboard] is not quite as big and physically lower than the *wenhua paizi* [culture bureau signboard]." In other provinces such as Guangdong and Guizhou, the CRB is a third-tier bureau (*xian ji ju*). In these cases, the CRB office is considered internally within the culture bureaucracy as a lower-ranking (*chu*-level) cultural relics *department* (CRD). Publicly, however, it is presented as a bureau (CRB) in a "one unit with two separate signboards" (*yitao ren liangkuaipaizi/ yige jigou liangkuai paizi*) relationship.[73]

There are also related cultural relics units, such as the Archeology Research Office, which, although partially managed (*fen guan*) by the CRD, is directly controlled by the higher-ranking CB. This office conducts the research over the actual effects of dam building on the relics in question. In some places, such as Guizhou, there is also an "archeology brigade," which, although based in the provincial capital, has the right to travel throughout the country to get involved in the discussions over particular projects.

Some provinces, such as Yunnan, only have CRBs and ABs at the provincial level.[74] In those provinces with bureaus below the province, at the prefecture level, there may be a cultural relics office (*wenwu ke*), and in some instances, like in Mianyang, Sichuan, there is a "three organs thrown together" relationship (*sange jigou fangzai yikuair*), in which the museums (*bowuguan lisuo*), culture (*wenhua ke*), and cultural relics are

---

72. Interview 04GY02, August 18, 2004; and Interview 05KM01, April 26, 2005.
73. Interview 04GY02, August 18, 2004; and Interview 05CD01, July 7, 2005. This is not unlike the internal divisions of the copyright bureaucracy. See Mertha, *Politics of Piracy*, ch. 4.
74. Interview 05KM01, April 26, 2004.

combined.[75] In addition, in some cases, there are cultural preservation units (*wenwu baohu danwei*) and national/historical preservation units (*guojia baohu danwei*) in lieu of these relationships, from the provincial level down to the county level. For example, in Zunyi, Guizhou province, such a unit is in charge of managing, supervising, and protecting the Zunyi Meeting Hall, where in January 1935, Mao returned from the political wilderness to take the helm of the CCP leadership. At the village and township level, there is only a "culture station," (*wenhua zhan*).[76]

GONGOs: The World Heritage Offices

China has left the management of its more than thirty World Heritage sites to local World Heritage offices. These offices are placed within the local government in which each World Heritage site resides. For example, Lijiang municipality's old town has its own World Heritage office within the municipal government. Yunnan also has a World Heritage office at the provincial level, because the Three Parallel Rivers region—which cover numerous counties and prefectures—also enjoys World Heritage designation.[77]

These offices are not outwardly political. But this can change with the circumstances surrounding particular policies. For example, the Dujiangyan World Heritage office was one of the key actors in reversing the policy decisions of the Yangliuhu project, as we shall see in chapter 4. The World Heritage office itself is an organization directly under the control of the Dujiangyan municipal government; it has consultative relations with its counterpart in Beijing, but it clearly seems to have a far more intimate—for which read substantive—relationship with the Dujiangyan government. Although the UNESCO office is situated within the "host" unit of the Ministry of Education, the Dujiangyan World Heritage Office is "independent" of the Education Bureau. It receives its administrative personnel and budgetary allocations (*bianzhi*) directly from and remains entrenched firmly within the Dujiangyan government.[78] The Dujiangyan World Heritage Office and its counterparts are best understood as government-operated nongovernmental organizations—GONGOs. That is to say, certainly in the case of Dujiangyan, it was as activist a unit as any of the other governmental or even nongovernmental hydropower opponents, but it was always completely in line with the preferences of the Dujiangyan municipal government.

75. Interview 05CD01, July 7, 2005.
76. Interview 04GY02, August 18, 2004.
77. Interview 04KM02, August 20, 2004.
78. Interview 04DJY01, August 5, 2004.

Other official bureaus oppose hydropower construction as a result of organizational interests. These are functional and nonpolitical, yet they somehow embody those "strange bedfellows" relationships only politics can provide, between these official units and the activists that exist outside of official channels.

The *Jiaotong* (Communication/Infrastructure): *Xitong*

There are several reasons why the bureaucracy charged with communication and infrastructure (*jiaotong xitong*) tends to be ambivalent about large-scale hydropower projects. First, hydropower can have a very direct and adverse effect on plans to develop waterways for shipping. As soon as hydropower stations are built, boats can no longer cross freely. They have to go through a series of locks, which can average up to two hours for each hydropower station—it takes four hours to go through the Three Gorges Dam. In addition, electricity usage patterns complicate the picture further. People use a lot of electricity during the day and less at night, which means that water levels are high during peak times of electricity use and are correspondingly low at night. This variation can affect the tonnage that can navigate waterways.[79] In general, this bureaucracy's promotion of or opposition to a project depends on the details of that specific project— if a dam project has a negative impact on road building, then they will oppose it. In some cases, this bureaucracy can display some internal conflict over different aspects of the dam project in question. For example, in the case of the Three Gorges, the Ministry of Communications was in favor of the upstream part of the project—that is, large-tonnage ships to Chongqing—but was quite concerned about the impact on transportation downriver from the dam. In sum, this bureaucracy chooses its battles carefully.[80]

Others

The construction (*jianshe*) *xitong* is a relatively hungry bureaucracy seeking to enhance its power by going after portfolios that may increase it.[81] The Ministry of Construction is in charge of geological and other types of parks, all of which are threatened by extensive damming.[82] This bureaucracy

---

79. Interview 04KM04, August 20, 2004.
80. Interview 04BJ01, August 2, 2004.
81. Ibid. On the Quality Technical Supervision Bureau in this context, see Mertha, *Politics of Piracy*, ch. 5.
82. Interview 04BJ05, August 4, 2004.

is responsible for the protection of *natural* heritage (*ziran yichan*), while the cultural relics bureaucracy, described below, handles cultural heritage issues. In Sichuan, the construction offices work extensively with the CRBs.[83]

The forestry bureaucracies (*linye xitong*) are charged with maintaining forests and, in this context, with controlling any erosion that might contribute to silting. On balance, they are against large-scale hydropower projects in the areas under review because of the substantial loss of wooded areas they would represent. The seismological bureaus are often against such projects because of the impact of reservoirs on the fault lines over which they extend.[84] The land resources (*guotu ziyuan*) office tends to be against reservoirs that submerge large tracts of land, especially when the land can be exploited for minerals and other resources. Like almost all bureaucracies, it is constantly seeking to increase its power.[85] Travel bureaus (*lüyou ju*) are also active, if weak, players, especially at Dujiangyan and at the upper reaches of the Nu, where eco-tourism provides an alternate source of revenue for local governments.

Often, these agencies cannot challenge the entrenched interests in favor of dam building directly. As a result, they leak information that the pro-dam units jealously guard, with the hope that the journalists will write about their side of the issue and thereby create more momentum in their favor.[86] As such, the impetus has to begin from outside the government in order

---

83. Interview 05CD01, July 7, 2005.

84. Interview 04BJ01, August 2, 2004; and Interview 04BJ05, August 4, 2004.

85. Ibid.

86. My participant observation notes from a meeting with a national level SEPA official and two journalists are instructive: "We met in a tea house. X [a senior official] was attending a meeting nearby. It was the first day of several and some of the participants had not yet arrived. This meeting was convened to discuss the Tiger Leaping Gorge controversy and was attended by people in various government offices, as well as a set of nongovernmental experts who were there to provide information. . . . X had taken leave from the meeting to come meet with us for 90 minutes or so, and when our meeting was over, X quickly went back to the official meeting (or the dinners that followed the meeting). . . . X started by talking about what had occurred during the meeting that day and what X had been arguing for. . . . While X was holding forth, A and B [the journalist and the NGO member, respectively] were furiously taking notes in longhand in their notebooks and X did not seem to mind. At one point, X started saying something and both of them closed their notebooks and stopped writing, just taking it in orally [this piece of information was beyond off the record, it was really on deep background]. Then, later, they started scribbling furiously again. After the first 30 or 40 minutes, A and B started to ask questions and to make suggestions, sometimes quite aggressively, and X would alternatively nod, debate a point, or expound upon what was being suggested. This very much appeared to be a situation of the give-and-take of ideas. A and B noted later that they were scribbling down information that was considered classified, and could be considered a state secret." Interview 05BJ02, July 4, 2005.

to provide the anti-dam units with a sufficient footing to challenge the pro-dam units. The opening salvo is "always made by these people,"—the media. There are two thousand newspapers and seven thousand periodicals in China, and it is simply impossible to control all of them. The Internet makes it even more difficult.[87]

The Role of Nongovernmental Organizations

The contours of China's contemporary state of affairs with regard to NGOs were best articulated by Dr. Jennifer L. Turner:

1) The Chinese government has opened political space for environmental protection activities, which has enabled an impressive growth in Chinese green NGOs and an increase in environmental activities by universities, research centers, journalists, and government-organized NGOs (GONGOs).
2) Independent Chinese environmental NGOs are at the forefront of civil society development in China.
3) Because environmental journalists enjoy more freedom in pursuing their stories than other beat reporters, they are quickly becoming a force pushing environmental awareness and investigations of local problems.
4) In the short term, expansion of green civil society in China is more dependent on improving organizational capacity of NGOs than an increase in political space.[88]

All of these points are supported by and consistent with the analysis to follow, with one important caveat: some of these NGOs are less patient than others in terms of trading political space for organizational capacity. In the following sections, I briefly trace this variation from the most risk-averse to the most risk-acceptant and speculate on some of the reasons why this variance exists in the first place.

87. Interview 04BJ05, August 4, 2004.
88. Jennifer L. Turner, "Clearing the Air: Human Rights and the Legal Dimension of China's Environmental Dilemma," statement to the Congressional/Executive Commission on China Issues Roundtable: "The Growing Role of Chinese Green NGOs and Environmental Journalists in China," January 27, 2003. For an excellent descriptive overview on Chinese NGOs and GONGOs, see Andreas Edele, "Non-Governmental Organizations in China," Programme on NGOs and Civil Society, Centre for Applied Studies in International Negotiations, Geneva, Switzerland, May 2005.

The Nature Conservancy

Among the more radical NGOs and activists in southwestern China, the Nature Conservancy (TNC) has the seemingly unenviable reputation not only of being "inappropriately" close to the Chinese government but also of actually functioning as if it were part of the Chinese government. Moreover, for the majority of dam and hydropower station opponents, TNC has dirtied its hands not only by working with "the government" but also specifically by working with the *Yunnan* provincial government, a major proponent of China's "anti-environmental" hydropower policy. Shorn of its sharp edges, such accusations are not completely untrue, but this criticism largely misses the point: close cooperation between TNC and the government is not regarded by TNC as a hindrance but rather as an integral part of its strategy in China. Indeed, the very existence of TNC as a foreign actor (the Chinese name for TNC—*Meiguo daziran baohu xiehui*—literally translates to "American nature conservancy association") severely limits its scope for maneuverability within the Chinese policy process.

The initial and continuing goal of TNC in China has been to turn vast parts of northwest Yunnan into a national park. This area has 3.1 million people and is the size of West Virginia, but it crisscrosses a complex set of jurisdictional and territorial boundaries. The short- to middle-term goal is to make Xianggelila ("Shangri-La," formerly Zhongdian) county into a national park (the "Yellowstone of China"). The Yunnan provincial government is even appealing to Beijing for help with this. Currently there is no National Park Law, and the current regulations regarding natural preserves are narrow in scope. TNC sees itself as in it for the long haul and is working for the long-term to set up such a model in northwest Yunnan so that it might be applied to other parts of China. One TNC official mused that such a model would encounter problems and mistakes at the beginning that could be ironed out with time—in any case, he continued, as modernity creeps in, there has to be a mix of modern and traditional. TNC sees itself as "a catalyst." Indeed, a vice director of SEPA told TNC that it need not worry about other parts of China—"you develop the model [in Yunnan] and we'll take care of applying it to other parts of China."[89]

As this implies, there is a considerable degree of cooperation between TNC and the Yunnan provincial government as well as with Beijing (although TNC is unique in that its main office is in the capital of Yunnan, Kunming, and not in Beijing). And TNC representatives are unashamed of

89. Interview 05KM05, July 26, 2005.

this approach. One put it this way: "China is not a blank sheet or an open playing field—it has an existing structure." So TNC feels that the most effective strategy is to work closely with government actors in China and others who may differ in terms of their values and even their desired outcomes but with whom some common ground can be established—"it is all about the process"—the classic "united front" strategy of the CCP. Yunnan in particular has an established governmental structure with which TNC must work; the goal is to integrate as much as possible into the existing decision-making mechanisms.

It was the Yunnan provincial government that invited TNC to examine biodiversity and ecosystems in northwestern Yunnan. TNC speaks directly with the Yunnan governor, provincial officials, and their counterparts in Beijing, Lijiang municipality, Nu prefecture, and elsewhere. TNC is in a very good position to raise issues and awareness within the government. It is often the only NGO invited to take part in government advisory councils and even some decision-making processes. Longtime TNC director, Rose Niu, was actually asked by the Yunnan provincial government to accompany them on a fact-finding trip to Xishuangbanna in southern Yunnan to investigate allegations surrounding a well-publicized logging controversy there.[90] In sum, TNC does not see itself as an opposition group; rather, TNC and the Yunnan government have come to see this partnership as mutually beneficial: TNC is able to do different things than the government can as regards the public inside and outside China, while TNC can accomplish many more of its goals in China with the government's help. Indeed, according to one TNC representative, many foreign NGOs have the wrong idea when it comes to strategy and tactics: the "shouting" approach may work in the United States, but it does not work well here, he argued.[91]

UNESCO

The United Nations Education Scientific and Cultural Organization (UNESCO) also finds itself in a somewhat sensitive position in China. On the one hand, it has overseen a gradual expansion of World Heritage sites in China to about three dozen. On the other hand, its representatives are extremely conscious of its subordinate status to the Chinese government in terms of leverage. In other words, while not necessarily satisfied with its current environment, UNESCO operates in a way fully consistent with an

---

90. Ibid. Many of the critiques leveled against TNC even attack Rose Niu, an ethnic Naxi, for not being a sufficient activist on behalf of "her people."
91. 05KM05, July 26, 2005.

acceptance of the status quo. Such behavior was quite evident during the Dujiangyan/Yangliuhu controversy.

During the Yangliuhu controversy analyzed in chapter 4, the role of UNESCO was marginal at best. Unwilling to incur political costs, UNESCO "outsourced" its objections to the Yangliuhu project to the Chinese media. UNESCO is understandably leery of being seen as "activist" in the eyes of the host government. It is acutely aware of its relative powerlessness. For example, a location must meet several criteria before it can be designated a World Heritage site, and these standards must continue to be met after the designation is conferred. If they are not, UNESCO can theoretically withdraw the designation. However, a UNESCO official was quick to add, "Politically, we cannot [withdraw the World Heritage designation]; we can only make suggestions."[92] These offices tend to tread extremely carefully, picking their fights—and more important, their tactics—with extreme caution. In the words of one knowledgeable source, "They do not do what they are paid to do."

On a recent trip to the Nujiang in region, the UNESCO delegation were accompanied by representatives from the Yunnan Provincial Research Center, the Provincial Construction Bureau, the provincial DRC, the Nujiang Prefecture Construction Bureau, the Forestry Bureau, the Environmental Protection Bureau, the Water Resources Bureau, and the Foreign Affairs Office. In addition to being "barbarian handled" by this entourage, local construction sites had been covered with cloth and survey ships had been removed from the area, according to one activist. Not surprisingly, UNESCO's assessment was positive: "The UNESCO delegation highly praised the efforts and contributions made to protect this area for the ethnic minorities living in the prefecture."[93]

That being said, UNESCO has taken a more active role in pursuing the Nu River and other cases. In 2003, UNESCO designated the Three Parallel Rivers region of northwestern Yunnan as a World Heritage Site. More recently, UNESCO has threatened to remove the Three Parallel Rivers from the list if the Chinese government continues with the original plans for the Nu River Project. At the same time, UNESCO is demanding more detailed plans for any project that is proposed for the Nu.[94] This approach represents a shift from UNESCO's traditional risk aversion; possibly this is because of

92. Interview 04BJ06, August 12, 2004.
93. Terry Wang, "Government Accused of Salween Hydropower Cover-Up during UNESCO Visit," *Interfax*, April 19, 2006.
94. "State of Conservation (Three Parallel Rivers of Yunnan Protected Areas)"/UNESCO, Decision 30 COM7 B. 11. See whc.unesco.org/en/decisions/1094, accessed July 29, 2007.

the appointment of a new chairman in October 2005, Zhang Xinsheng, who had been vice minister of education. Although it is important not to inflate expectations of UNESCO's ability to leverage Beijing, it does appear that UNESCO is taking advantage, albeit modestly, of the political spaces emerging from this new environmental activism in China.

Friends of Nature

Friends of Nature (FON), the first environmental NGO in China, was founded in 1994 by Liang Congjie, an environmental activist from a reform-oriented lineage that goes back to the Qing Dynasty.[95] Liang's grandfather was none other than Liang Qichao, a prime mover of the 1898 Reform Movement. Liang's father was Liang Sicheng, the Beijing city planner after 1949. Liang Sicheng believed that the walled inner city of Beijing should be preserved but was rebuffed by the Soviet experts and fell out of favor with Mao Zedong; he died, officially in disgrace, during the Cultural Revolution.

In 1993, Liang Congjie decided with a group of friends that forming an NGO would be the most effective way to expand public awareness about the environment. After "a long and difficult year of meetings," FON was officially registered in December 1994 as the Academy for Green Culture, finding a home within the Academy for Chinese Culture. It does not see itself as a lobbying group or as an information regulatory agency. Yet its past successes and current work belie its modest self-image. Indeed, in a significant shift of the government-NGO roles in the policy process, Liang Congjie (as well as some others) was instrumental in pressuring the Chinese government to include the State Environmental Protection Administration as a participating agency in the "Develop the West" campaign.[96]

FON and other groups like it recognize that there is very limited space between state and society within which they can undertake effective action and hope for substantive results.[97] One of the ways that has proven most successful is the dissemination of ideas and images to the broader public as a way of "framing" the national debate. Indeed, much of the power of China's NGOs comes from the media. NGOs uncover wrongdoing by the

95. For a fascinating survey of these three generations of Liangs, see www.rmaf.org.ph/Awardees/Biography/BiographyLiangCon.htm, accessed July 28, 2007.

96. See also Noah Bessoff, "One Quiet Step at a Time," *Beijing Scene* web edition, www.beijingscene.com/v07i010/feature.html, accessed July 29, 2007.

97. In August 2004, FON helped found the China River Network with six other NGOs: Global Village of Beijing, Green Earth Volunteers, Institute of Environment, Wild China Film, Brooks Education Institute, and Green Watershed. They banded together in part because they felt that they could achieve more as a group than as discrete NGOs.

government and corporations and transmit such information to the media. In 1996, FON's snub-nosed monkey campaign in Yunnan was responsible for the first NGO-initiated national policy change in Chinese history.

In the early 1990s, local officials in Deqin county, one of the poorest places in China, with an annual per capita income of fifty U.S. dollars, switched their logging operations to a tract of forest adjacent to the Baimai Xueshan nature preserve, home to two hundred snub-nosed monkeys—one-fifth of the entire population of the species. Xi Zhinong, a photographer attached to the Yunnan Provincial Forestry Bureau, was informed by the director of the local forestry bureau that the logging would proceed unless another source of revenue for the county's annual budget could be located. One of China's leading environmentalists, Tang Xiyang, suggested that Xi write a letter to State Councilor Song Jian. Sympathetic to environmental issues, Song ordered a stop to the logging. After this was reported in the press, two investigation teams from the forestry bureau were sent to the area, and the central government granted a special subsidy to the county, stopping the logging temporarily.[98]

The success of FON demonstrates how important it is that many NGO leaders have former (and even ongoing) careers in publishing, media, and communications as well as with sympathetic officials within the government.

Green Watershed

On the extreme end of the spectrum of viable Chinese NGOs is Green Watershed. Founded in 2002, it is the first Chinese NGO that specializes in water management.[99] Its founder and director, Yu Xiaogang, has worked for years on watershed and related issues within the Three Parallel Rivers region of northwestern Yunnan. Interestingly, Yu is a member of the Chinese Communist Party, yet he holds no government post. His colorful past provides him with Chinese socialist bona fides that make it

---

98. Interview 05BJ04, July 6, 2005. Of course, this was not the end of the story. According to the U.S. Embassy, "In spring 1998, Xi learned that the logging company was still operating in the forest. He took a crew to film the operations secretly with the result that the county governor had to confess his mistakes. In Deqin, the county vice-governor has since been demoted. Officials there may hope they never see Xi again and, indeed, there are rumors of threats against him. That summer, China was hit by devastating floods that were largely blamed on deforestation in river catchments." See www.usembassy-china.org.cn/sandt/webmonk.htm, accessed May 16, 2007.

99. Rose Tang, "The Way China Is Managing Its Watersheds Is a Matter of Life and Death, Says Yu Xiaogang, Long-Time Communist Party Member and Founder of the First Chinese NGO Specializing in River Management," *Hong Kong Standard*, April 16, 2005.

very difficult for authorities to marginalize him or his message. In 1968, Yu volunteered to aid the Burmese Communists in their fight to overthrow the government in Yangon. His current work is largely seen as an effort to help Chinese peasants, which had been the backbone of the CCP victory in 1949. Indeed, the full name of Green Watershed is the *Yunnan sheng dazhong liuhe liuyu guanli yanjiu he tuiguang zhongxin,* the "Yunnan Center for Watershed Management, Research and Promotion by the Masses."

Green Watershed has been extremely active in the ongoing Nu River controversy as well as in many other equally broad and expansive projects. Yu was responsible, among other things, for sponsoring a trip to the Manwan Dam site for Nu River Valley peasants and local officials to show them how their counterparts displaced by the Manwan Dam were reduced to collecting garbage for a living. After filming the event, the group "burned" DVDs for distribution among the villagers in Yunnan. In addition to his unshakable revolutionary credentials, Yu has been very careful about staying within the boundaries of the law, knowing that just as he has used the law as a way to leverage his agenda, so too can officials use the law to stop him. Moreover, Yu has had his home broken into, and his staff has received threats; he negotiates regularly with insurance companies to make sure that members of his staff are fully covered in the event of some "misadventure." Like the liberalization of the media, the NGOs' establishment of an autonomous or semi-autonomous space is not a linear evolutionary process; it is more dialectical in nature. Just a month after my own field research in July 2005, during which I enjoyed largely unfettered access, Michael Büsgen embarked on fieldwork for his master's thesis. He later described the political environment he encountered. It is worth quoting him at length:

> The different dynamics and factors which shape the spaces for social organizations in China became very apparent during my visit to Beijing and Kunming for…research in July and August 2005. This time coincided with what many local NGO activists described to me as the most difficult period in the history of grassroots NGOs in China. Almost all of the NGOs, which I met during that month, had in the previous weeks been visited by State Security officials, who inquired about their funding sources and the nature and objectives of their work. All organizations who previously had failed to register under the MOCA [Ministry of Civil Affairs], and therefore had no legal status, were asked to reregister—even though the possibility of getting official approval appeared fainter than ever.[100]

100. Büsgen, *NGOs and the Search for Chinese Civil Society,* 3.

Having sketched out the actors and interests, I devote the next three chapters to the political process of hydropower policy, paying special attention to the role of actors traditionally kept out of the policymaking process—that is, the anti-hydropower forces detailed above. Taken together, these cases suggest that the political process is indeed becoming more pluralistic in ways that are not represented in the existing literature.

3 | From Policy Conflict to Political Showdown: The Failure at Pubugou

Resettlement is the most difficult thing in the world.
—Hanyuan County Deputy Party Secretary Bai Rangao, 2004

Pubugou, I cannot talk about that....
—NGO Officer, Beijing, March 2006

In many ways, the fall of 2004 was like any other in sleepy Hanyuan county, Sichuan province: crop harvests, shorter days, and colder weather. But in other respects, that autumn in Hanyuan was unique: up to a hundred thousand people demonstrated against the imminent groundbreaking of the Pubugou Dam and against acquiescence of local governments to the resettlement of almost half of Hanyuan's total population.

The protests occurred over the course of several weeks, with clashes between peasants and police and the destruction of vehicles—several were pushed into the river—and other state property. On October 28, 2004, protesters actually occupied the dam itself. These demonstrations culminated in the detention of the party secretary of Sichuan province for several hours by an angry mob before he was able to escape through a back door.[1] The Pubugou protests were said to be the biggest since 1989 and reportedly the largest rural protests since the founding of the People's Republic. It is therefore ironic, even baffling, that these protests had absolutely no effect on the dam project aside from delaying the opening of the floodgates of the diversion channel.

It was the nature of the opposition—the very form it took—that made it impossible to expand the sphere. Although peasant dissatisfaction had been simmering for a long time, it was diffuse and only very loosely organized. And apart from a few de facto leaders among the peasants, there were no visible allies within the local government to act as policy entrepreneurs who

---

1. See, for example, Jason Leow, "Official Rescued as China Dam Protest Eases: Ten Thousand Troops Have Been Sent to Quell Unrest that Flared in Sichuan Province over Construction of the Pubugou Dam," *Straits Times*, November 8, 2004.

might have instigated policy change in a way that could have led to a different outcome. In addition to this dearth of policy entrepreneurship and the corresponding absence of a meaningful coalition, it was the government and not the opposition that successfully altered the media image, strategically changing the issue frame from one of economic development to that of political crisis. Because of the size and the nature of the protests, it was impossible for the protesters to mobilize the media and NGOs or to elicit the support of sympathetic local officials—the situation had become too political. It was no longer an issue of economic development, environmental concerns, or even compensation issues. Rather, it had become one of social stability—the one issue frame that unambiguously trumps even economic development. The media outlets were unable to report on the events except afterward, once the outcome was a fait accompli. Even then, it could only do so under strictest standards of state censorship. Thus, it was impossible to mobilize even a small coalition to oppose this project in an effective fashion. The press was kept out of the area, and local opponents within the government as well as NGO activists were frightened into silence and inaction.

Ultimately, while there were some minor political changes immediately following the protests—including the dismissal and arrest of local cadres accused of corruption as well as a campaign to reestablish stability in the affected areas—at the end of the day, there was next to no impact on policy. The outcome—as distinct from the process—of Pubugou is similar to earlier dam project construction in China while substantially different than the cases explored in chapters 4 and 5.

Background

Despite, or perhaps because of, its remoteness, Hanyuan county is close to some of the more important historical accidents that changed the course of modern Chinese history, the most recent of which was the crossing of the Dadu River at Luding by the Chinese Red Army in 1935:

> Had the Red Army failed there, quite possibly it would have been exterminated. The historic precedent for such a fate already existed. On the banks of the remote [Dadu] the heroes of the Three Kingdoms and many warriors since then had met defeat, and in these same gorges, the last of the T'ai-p'ing rebels, an army of 100,000 led by Prince Shih Ta-k'ai, was in the nineteenth century surrounded and completely destroyed by the Manchu forces under the famous Tseng Kuo-fan.[2]

2. Edgar Snow, *Red Star over China*, New York: Random House, 1968, 194.

Keenly aware of Prince Shih's fate, which was the result of an avoidable delay (a three-day celebration of the birth of the prince's son), the Chinese Communists during the Long March in 1935 raced along the southern banks of the river to reach the Luding Bridge and cross the Dadu River before Kuomintang (KMT) reinforcements arrived. After an extended forced march, they arrived at the bridge and, at the cost of a dozen or so volunteers who pulled themselves across the massive chains that held the bridge—the KMT had removed planks that constituted the bridge itself— overcame the machine gun emplacements on the other side. Over the next three days, the bulk of the Red Army crossed the bridge and eventually went on to establish their camp in Yan'an, to the north, the following year. Since the Luding Bridge was the most westward point that the communists could cross the river, their exploits at Luding arguably changed the course of twentieth-century China.[3]

Despite this historical pedigree, Hanyuan and the surrounding areas are for the most part calm and quiet places, like most of China's rural farmland. That all changed in late 2004.

Breaking Ground

Hanyuan is one of the key sites where China is attempting to implement its "Develop the West" policy, specifically to harness the Dadu River's immense hydropower potential. The dam at Pubugou is the largest of an enormous project to build seventeen large-scale dams, in addition to 356 smaller power plants, along the Dadu.[4] Power generation is expected to start in 2008, and the entire project is slated to be completed in 2010. The dam will be almost six hundred feet high. When completed, it will house China's fifth-largest hydropower station.[5] The average power generation rate is estimated to be 146 billion kilowatt-hours. The reservoir will flood 44,383 *mu* of land.[6] The annual national tax revenue is estimated to be 7 billion RMB, while the local tax revenue for the county is estimated to be 1.7 billion RMB. The predicted rise in local GDP is estimated at 20 percent, but there is some dispute over the distribution of revenue. Because the Pubugou hydropower station is registered in Chengdu and not Hanyuan, Chengdu will receive an operation tax, which is submitted to the locale in which the

3. Snow's account has been recently challenged by Chang Jung and Jon Halliday in *Mao: The Untold Story*, New York: Knopf, 2005.

4. Shi Jiangtao, "Landslides Pose a Bigger Threat," *South China Morning Post,* January 4, 2005.

5. Shi, "Exodus Forced by Dam Under Way," *South China Morning Post,* December 13, 2005.

6. Tan Xinpeng, *Dadu he shuidian yimin ju'e liyi liushui* [Enormous Hydropower Compensation Benefits of Dadu River *Yimin* Washed Away], *China Youth Daily,* October 27, 2004. One *mu* is equal to 0.165 acres.

project is registered. A development tax is shared by the national, provincial, and local governments farther down the administrative hierarchy, with the latter getting a rather small piece of the pie.[7]

The costs, however, are also enormous. An estimated 100,000 to 120,000 people—around a third of Hanyuan's entire population of 310,000—have been or will be displaced; 33,000 were moved out of Hanyuan county altogether in the summer of 2006. Although some people had been resettled as early as November 2001 and the first diversion tunnel was under construction by October 2002, the feasibility report was not submitted by the National Development and Reform Commission (NDRC) until July 2003; it received State Council approval the following spring.[8] The Ya'an municipal and Hanyuan county governments were both somewhat in favor of the project—privately, some local officials remained far more ambivalent—but by far the most enthusiastic government entity was the Sichuan provincial leadership.[9]

In addition to the human and other social costs, the economic opportunity cost of the dam project is also quite significant. The land surrounding the reservoir is relatively flat and extremely nutrient-rich—the area is known for its fruit and its outstanding *huajiao* ("numbing peppercorn"). Hanyuan enjoys good weather (with no frost for eleven months of the year, it has three harvests annually), and even though it has experienced some flooding, it has largely been contained and controlled by a dense network of dykes and diversion canals. The reservoir would submerge 14 percent of Hanyuan's farmland.[10]

Yet it is for precisely this reason that the site was selected. Pubugou was chosen to be the site for the largest dam of the Dadu River Hydropower Project because the banks of the usually quite narrow Dadu River tend to be fairly wide around the Pubugou area, allowing for a sizable reservoir. This factor trumped the geological instability of the area, which makes it prone to landslides.[11] There were plans for a dam project around the Pubugou area as early as 1952, but they were scrapped because it would have been bad for the local farmers right when the newly established People's Republic was implementing policies to benefit the social class on whose backs the Revolution had been fought and won—the peasants. Under the orders of Mao Zedong, the project was put on hold indefinitely.[12]

7. Ibid.
8. This was not illegal or even irregular because the project is largely confined within the province, but it does seem odd that a request for approval was sent to Beijing at such a late stage of the construction and planning process. Pubugou Interviewee, July 8, 2005.
9. Ibid.
10. Tan, *Dadu he shuidian yimin ju'e liyi liushui.*
11. Shi, "Landslides Pose a Bigger Threat."
12. Pubugou Interviewee, July 8, 2005.

The project was revived after 1978 by the Sichuan provincial government, and plans to implement it were announced in the mid-1980s, although the groundbreaking did not begin until 2000. In March of that year, the Dadu River Hydropower Corporation (*Dadu he shuili gongsi*) was established and officially registered, and in November the Guodian Dadu River Hydropower Corporation (*Guodian Dadu he shuili gongsi*) was established. Construction was to be handled by Guodian, while extra personnel were to be provided by Hanyuan county.[13] In December, the Sichuan Provincial Government Planning Commission (precursor to the Sichuan DRC) and the Guodian Dadu River Hydropower Corporation submitted the Pubugou project proposal to their counterparts at the national level.[14]

The Economics of Resettlement Compensation

Controversy surrounded the project from the very beginning. The issues of resettlement and related compensation were very complicated. Opinions varied over what the affected people would be paid, where they would be resettled, and who would share the profits stemming from hydropower generation. In April 2001, the Sichuan provincial government sent a letter to the State Development Planning Commission (which became the National Development Reform Commission in 2003) promising that "the Sichuan provincial government will take responsibility over the Pubugou power station relocation.... Expenses will be used by the Sichuan provincial government according to the estimation approved by the Central Government."[15] But this commitment was vague and somewhat conditional.

The main problem was establishing a standard for adequate compensation. But administratively and politically—not to mention economically (from the standpoint of the hydropower companies and the government recipients of the profits)—it was quite difficult to establish a standard that conformed to reality. This was because, given the rich land and good weather, peasants in Hanyuan county were actually doing quite well and enjoyed a relatively high standard of living: "Farmland here is so fertile that in one year we can grow enough to last us three years," said one villager.[16] However, for the Sichuan government to justify the project on

13. This is in some dispute, as some argue that local people refuse to be employed by Guodian. If this is the case, there is very little that Hanyuan can do because it is only a county, while Guodian enjoys de facto provincial-level rank. Tan, *Dadu he shuidian yimin ju'e liyi liushui.*

14. Pubugou Interviewee, July 8, 2005.

15. Ibid.

16. Shi Jiangtao, "Peasants in Upstream Fight to Halt Dam," *South China Morning Post,* January 5, 2005.

Figure 3.1 Construction of the Pubugou Dam. Photograph by the author, August 2006.

economic development—that is, policy—grounds, it had to demonstrate that the peasants were poor.[17] As a result, there were powerful incentives to push compensation levels downward. It is estimated that the difference between the compensation package offered—based on fourteen-year-old standards[18]—and contemporary national standards for compensation is in the neighborhood of 1.2 billion RMB (US$143 million).[19] As this would entail a substantial reduction in operating costs, Guodian was happy to oblige,[20] labeling the Hanyuan reservoir area as one of "deep valleys and

17. Tan, *Dadu he shuidian yimin ju'e liyi liushui.*

18. "Massive Protest by Sichuan Farmers Squashed by Police," *Epoch Times,* November 10, 2004.

19. "Outrage Emerges over Pubugou Hydropower Project in Sichuan: Residents Forced to Move to Make Way for a Dam on the Dadu River Are Angry about the Meagre Compensation Offered," *Interfax,* October 28, 2004.

20. Indeed, this became a self-fulfilling prophecy as the Sichuan provincial government announced in March 2001 that all non-dam-related development was to stop in Hanyuan. To drive the point home, the Hanyuan party secretary announced that anybody who opposed the

high [barren] mountains," not the more accurate designation of "high yield farmland," during the approval process for Pubugou power station.[21]

On October 25, 2002, the provincial government issued a document that defined the Pubugou project relocation policy and compensation standard. In March 2003, the Pubugou relocation plan was approved by the Water Resources Planning Institute. On September 22, 2003, a meeting was held in the provincial capital of Chengdu by the Sichuan provincial government to establish a division of labor and assign specific tasks to various local government agencies.

More than three years after the resettlements had already been underway, on May 19, 2004, the Sichuan Provincial Government Large Hydropower Station Relocation Office issued a notice regarding the compensation standard for Pubugou.[22] This information was disseminated to Hanyuan cadres in June. Houses with a frame structure had a compensation level of 473 RMB per square meter, brick and concrete houses received 300 RMB per square meter, and brick and wood houses received compensation levels of 180 RMB per square meter. Not only were these standards lower than market prices had been in the past, they were far lower than the prices being charged to subcontractors that were building homes for the construction camp at Pubugou, which ranged from 520 to 680 RMB per square meter.[23]

In order to justify such low compensation rates, supplementary materials, such as the "Relocation Questions and Answers," stipulated that the relocation should follow the "Large and Medium Size Hydropower Station Projects Condemnation and Relocation Regulations" issued on May 1, 1991. But the resettlers (*yimin*) objected to a fourteen-year-old standard for compensation. According to villager Song Yuanqing, the political negotiation representative for Hanyuan county, in 1991 a village cadre's income subsidy was 7 RMB; in 2003 it was 50 RMB. In 1991, a bus ticket from Fulin to Wusi cost 1.2 RMB; in 2004 it cost 9 RMB. In 1991 Hanyuan had tax revenues of 3 million RMB from local businesses; in 2003 the figure had risen

---

Pubugou project was also personally opposing the party secretary himself. As a result, from April 2001 on, all infrastructure construction not related to the dam was stopped. Because the builders of the power station did not want participation from local companies—the power company is registered in Chengdu Xigao district and the construction camp in Ganluo county, Liangshan prefecture—almost all of the local construction companies went out of business. Pubugou Interviewee, July 8, 2005.

21. Tan, *Dadu he shuidian yimin ju'e liyi liushui;* and Christopher Bodeen, "One Killed in Mass Protest over Dam in Western China," *Associated Press,* November 1, 2004. This is far from an uncommon practice; see Shi, "Peasants in Upstream Fight to Halt Dam."

22. Although possibly illegal, this is not uncommon. This also occurred at the Xiluodu hydropower station along the Jinsha River. Pubugou Interviewee, July 8, 2005.

23. Ibid.

to 40 million RMB. These issues were left hanging, as peasants were being readied for resettlement.[24]

Resistance to Resettlement

As noted, even though resettlement began as early as 2000, official compensation would have to wait, as the Hanyuan authorities had not yet received proper authorization. The *yimin* were given money only for immediate and pressing costs, and the balance would be delayed until the resolution of the standards issue. The stopgap measure was articulated as follows: "We'll undertake relocation first, and when the [official] relocation policy is established, it will be applied [retroactively] to those who have already been relocated."[25]

In 2001, the first people were being moved out of Sanguzhuang village, Sunhe township, primarily an Yi minority village. Rather than moving to the upper reaches of the area in which they had lived,[26] they were moved to Qianyu, into a village of new but very shoddy houses with no attached land.[27] They resisted, and their opposition concentrated on the absence of any official policy from which they could draw even a minimum of comfort. According to one Yi farmer:

> In August 2001, we signed the relocation agreement with the Shunhe government and acceded to the requests of the county government. At that time, the government promised to construct water supply, highways, electricity, and also open new farmland for the relocation site. But in April 2002, the County Relocation Office suddenly changed the original agreement, and moved us to a barren slope in Qianyu prefecture, Jiuxiang district, instead of moving us upland to the original site in Erzichang.[28]

As a harbinger of things to come, their resistance led to the mobilization of the People's Armed Police with twenty trucks to "assist" with the relocation as well as the incarceration of ten villagers until they signed an agreement to relocate. In one skirmish with the police, another ten or so villagers were wounded, some quite seriously. Their defiance did not end with their forcible relocation. Once they were resettled, they moved themselves to higher ground, to the top of Erzichang hill, because there was some attached land (albeit inferior to the land they had been forced to relinquish) and it was a

24. Tan, *Dadu he shuidian yimin ju'e liyi liushui.*
25. Pubugou Interviewee, July 8, 2005.
26. The reason for the change was that the Chengdu Survey Institute wanted to build a timber mill at this site, so the agreement was changed, unbeknownst to the *yimin* who would be affected. Tan, *Dadu he shuidian yimin ju'e liyi liushui.*
27. Pubugou Interviewee, July 8, 2005.
28. Tan, *Dadu he shuidian yimin ju'e liyi liushu.*

little closer to their original village of Sunhe. They did so without governmental approval. Although this area is very cold in the winter and rainy in the summer, in 2005 there were as many as two hundred people still living there, in makeshift housing they had fashioned themselves.

On July 4, 2002, the inhabitants of Jiaotuo village were beaten after refusing to move. Several were also imprisoned and were only set free after they signed an agreement that they would accept their resettlement to another village, Qianyu.[29]

By 2004, most of the people had yet to be resettled. One of the emerging problems was a resettlement bottleneck: there were many *yimin* but little land in the area on which they could be resettled. The Hanyuan county government sponsored trips for the local *yimin* to see the places to which they would be moved. These included Wutongqiao, Leshan; Mianyang municipality; and Minshan county of Ya'an municipality. However, because these places had comparatively poor land and a very underdeveloped commercial infrastructure, these trips made the people less willing to move, the opposite of the effect that the trips were intended to achieve.[30]

In addition to the farmers, there were also urban citizens (*jumin*) of the Hanyuan county seat who were to be resettled, but they objected to being allotted the same compensation of 300 to 400 RMB per square foot of property, as had been doled out to the peasants.[31] The urban residents argued that their property was worth far more and that their compensation levels should reflect this. Moreover, the buildings that were built in the resettled areas were of poor quality, but they were being sold to them by the local governments at a price that was higher per square foot than what they had received in compensation.[32]

"Drawing Water from the Sky": The Case of Dashu

The fate of Dashu provides a particularly good illustration of the issues at stake from the point of view of the villagers. Dashu is a township (*zhen*) and former people's commune that was established in 1955 by carving space out of the surrounding forests.[33] As early as 1952, levies were built along

---

29. Ibid.
30. Pubugou Interviewee, July 8, 2005.
31. Laid-off (*xiagang*) workers who had been able to survive by renting out their houses to businesses were faced with the loss of their last valuable asset. Tan, *Dadu he shuidian yimin ju'e liyi liushui.*
32. Pubugou Interviewee, July 8, 2005.
33. Hanyuan had established twenty-eight people's communes (*renmin gongshe*) by September 22, 1958. *Hanyuan xian zhi* [*Hanyuan County Annals*], Chengdu, Sichuan kexue jishu chubanshe, 1994, 23.

Figure 3.2 Three views of Dashu—above, at, and below the proposed waterline. The top left photograph is of the area of Dashu township that will remain above the water level. The bottom left photograph is of a sign that marks the "normal reservoir level" (*zheng chang xushuiwei*). The Chinese character on the front of the condemned building above is *chai*, which means "tear down, dismantle." Photographs by the author, August 2006.

the Dadu River to prevent flooding into the area that would eventually become Dashu township. In the following decades, Dashu residents expended a tremendous amount of energy and resources to build a dense network of irrigation channels, small sluice gates, and levees to control the flooding through irrigation channels. They had also reclaimed 4,000 *mu* from what had been river areas and opened up 9,000 *mu* of farmland, all told. The citizens of Dashu even built a bridge across the Dadu River, which I crossed on a visit in August 2006. These projects required a great deal of sacrifice, and a few people—some say as many as several dozen—even died in the process.[34]

In the early 1970s, "*nongye xue Dazhai*" ("in agriculture, learn from Dazhai," a well-known commune) had become a national slogan, and Dazhai itself had become a model to be emulated by countless villages all over

34. Pubugou Interviewee, July 8, 2005.

China. But in Sichuan, long before Dazhai gained such prominence, it was just as likely to be "*nongye xue Dashu*" ("in agriculture, learn from *Dashu*)."[35] At the height of Dashu's popularity, thousands of pilgrims reportedly visited the township every day. Dashu's own slogan was that it was "drawing water from the sky" (*kong zhong yin shui*). These efforts in developing Dashu into a viable agricultural and commercial center were redoubled as a result of Party Secretary Zhao Ziyang's efforts to break up the people's communes and encourage rural economic experimentation.[36]

By the reform era, the hard work of Dashu's peasants had paid off. The farmland around one of its nine villages, Hailuo, was yielding 1,700 kg of rice per *mu*. Garlic was being grown as a cash crop; able to net 4,000 RMB per *mu*, it was exported to Japan and to Southeast Asia. And if this were not enough, 500-pound pigs slaughtered around the Lunar New Year festival were not uncommon. In Hailuo village itself, there were new houses, a supermarket, a karaoke parlor, and other entertainment facilities. Local cadres boasted that each family owned at least a motorcycle and a color television and that many families owned cars.

But the Pubugou project will submerge most of Dashu's nine attached villages (*cun*). As much as 100,000 *mu* of prime farmland will be submerged, and an estimated 18,000 Dashu villagers will be resettled.[37] The people who had been active creating Dashu's present wealth, many of them now in their sixties, were especially unwilling to move. Yang Jijin, the former Party secretary of Maipin, one of Dashu's villages, expressed this unwillingness thus: "we removed stones and sand from the earth.... it took forty years to create such arable land and we suffered a great deal during that time."[38] Residents demanded stipulations that those who had been killed in the construction of the Dashu irrigation network during the 1960s and 1970s be compensated as well. Understanding this all too well, certain local officials (even some at the county level) sought a compromise. Time and again, they were overruled by the higher-ranked cadres in Hanyuan and beyond.

Memorializing the Situation

Some officials from the Hanyuan county government and elsewhere did have considerable empathy for the *yimin* ("There but for the grace of the

35. *Hanyuan xian zhi*, 26, 879 (my emphasis).
36. Playing on his name, a local saying had it, "If you want to eat, look for [that is, *zhao*] Ziyang." Gady A. Epstein, "Proud of Protest, Silent on Politics," *Baltimore Sun*, February 13, 2005.
37. Pubugou Interviewee, July 8, 2005.
38. Tan, *Dadu he shuidian yimin ju'e liyi liushui*.

Party go I"). Indeed, cadres within the Hanyuan county government expressed their own reservations over the enterprise. Nevertheless, they not only faced demotion or dismissal in the event of noncompliance, they also faced legal action or worse.[39]

The situation was made even more difficult because the *yimin* had studied national laws and regulations, thus forcing local cadres to do the same if they were to have any hope of persuading the resettlers. In doing so, the Hanyuan authorities discovered that the compensation standard used in the "Preliminary Design Report" diverged from the national land standards and from the experience of comparably sized hydropower stations (the total shortfall has been estimated at 11.9 billion RMB). In addition to the compensation standard issues described earlier, forested land was also undercompensated, while compensation for green seedlings was not included in the total estimates. The properties to be left behind but above the water line were not included in the original compensation package, and neither were properties needed at the relocation site to make the land habitable. As to the relocation of enterprises, workers' lost wages, unemployment subsidies, and social security payments were all excluded from compensation estimation. Finally, a listing of small processing enterprises in villages that ran to 297 households excluded 123 households.

As a result, sympathetic local officials proposed a meeting with county people's representatives (*xian renmin daibiao*), resettled farmers' representatives (*yimin daibiao*), and representatives of the city dwellers to be resettled (*jumin daibiao*). But the meeting never took place. According to a knowledgeable source, the officials who sympathized with the *yimin* and the *jumin* may well have been in the majority, but they were overruled by ranking cadres who were not as closely tied to the region. The *yimin* wrote scores of "letters of opinion" (*yijian shu*) to express their dissatisfaction with the situation, culminating in a petition with twenty thousand signatures. In a shift that represented a change in the *yimin* demands that continued throughout the remainder of the crisis, they shifted their stipulations away from compensation and coalesced around petitioning that the height of the dam be lowered so that they would not have to move from some of the richest and most populous sections of the area to be flooded.

They petitioned the Hanyuan county government. Not surprisingly, they were rebuffed. They took the letters to the Ya'an municipal government but were snubbed again. Then they took them to the Sichuan provincial government—to the Party committee, the Sichuan DRC, and to the Resettlement

---

39. Pubugou Interviewee, July 8, 2005.

Bureau (*yimin ju*)—and were rebuffed by each in turn. Finally, in two trips in July and August 2004, a group of representatives took these letters all the way to Beijing. They presented their case to the State Council, the NDRC, the Ministry of Land Resources, and the Ministry of Agriculture. A somewhat lower-ranking department head (*chuzhang*) in the NDRC who met with the *yimin* representatives said that this was an important issue and that he would take it up with technical experts to see if a resolution could be achieved. Upon hearing this, the representatives returned to Hanyuan with the understanding that their case was being handled properly.

A month later, the *yimin* representatives who had gone to Beijing again contacted the NDRC to inquire about the status of their petition. Their interlocutor responded by saying that it had been looked into but that the decision had not yet been made by the Sichuan provincial government. The *yimin* were told that they should prepare for the resettlement and that they should obey the orders of the provincial government. The *yimin* representatives replied that if their demands were not taken into account, they would disrupt the scheduled shift to the diversion channels (*tong shui*) planned for November 2004.

By this time—after the signature campaign and the trips to Beijing—the Hanyuan county government was increasingly aware of the degree of the people's discontent. From this point onward, tensions began to escalate.[40]

The Pubugou Protests of 2004

On September 21, 2004, earthmovers descended on Wangong village to acquire soil to use in the dam construction, but they were prevented from doing so by a group of villagers. This was the first time that any of Hanyuan's villages had actually been "breached" by Guodian. Guodian responded by bringing a hundred or so reinforcements to start the earthmoving operations, but they were met by twenty thousand protesters. Guodian was forced to stop. This became known as the "September 21 Incident" (*921 shijian*). The Hanyuan county government mobilized local cadres throughout the area to confront the protesters and to reiterate that the dam project was "a done deal," that the protests had become meaningless by this point, and that they must obey the government. However, the protesters did not budge, and Guodian was not able to begin the earthmoving operations. In Maoist terminology, this was the "single spark" that began the 2004 Hanyuan "prairie fire."[41]

40. Ibid.
41. Ibid.

One month later, on October 20, the Sichuan Party Secretary Zhang Xuezhong, and deputy Party secretaries Li Congxi and Gan Daoming of the Sichuan Party Committee as well as the deputy provincial governor, Zhang Zuoha, went to Hanyuan for the purpose of surveying the Pubugou engineering project. It became evident that they had no plans to meet with any of the protesters. Zhang made it known that the provincial Party committee, the provincial government, and officials from Liangshan prefecture and Ya'an municipality should fully support the Pubugou project. He went on to praise Guodian for "overcoming difficulties in building the Pubugou project in such a complicated geological setting." Regarding resettlement, Zhang requested that cadres from the provincial, prefecture, and municipal levels offer whatever help they could to the Guodian Dadu River Corporation, yet he remained mute concerning the *yimin*. One of Zhang's themes was the notion that this project, along with all the challenges emanating from it, was not simply the problem of Guodian; it was also a problem for the Sichuan government. By thus expressing "ownership" of the issue, the implication was clear that whoever opposed the Pubugou project would also be opposing the Sichuan government. Zhang concluded the meeting by asserting that he would return to write the obligatory calligraphy—"develop clean energy sources, realize sustainable development"—at the November ceremonies for the start of the dam construction.[42]

Exactly one week later came the "day of reckoning," the scheduled date for diverting the water. Some of the *yimin*, well aware of this date, undertook a twenty-kilometer trek to the site on foot. Because it had gotten so cold, only a few thousand people stayed at the site.

On October 28, the crowd of protesters got much larger, from approximately fifty to one hundred thousand people—estimates differ—and the group proceeded to occupy the dam site. What is even more remarkable is that this demonstration was not organized, except in the loosest possible fashion. According to one observer, what united these people was their opposition to the project; there was no exogenous organization involved. The police cordoned off the area, afraid that the protesters were going to descend on the general headquarters of Guodian, but the protesters had plans only to occupy the dam.

It was inevitable that, since the crowd had been milling around for some time, some people had to relieve themselves. An old woman who needed to go to the bathroom tried to break through the line and enter what turned out to be a restricted area to do so. She was subsequently beaten by the

42. Ibid.

police. On seeing this, the crowd attacked. They set several cranes on fire and overturned two police minivans and pushed them into the Dadu River. That night, those who stayed behind in the villages sent food to the protesters, but one of the people bringing the food on a motorcycle was killed in a traffic accident.[43] Protesters brought the body to the Hanyuan government compound.

On the next day, October 29, as a sign of protest, all of the stores in Hanyuan shut their doors. Urban middle school children, protesting on behalf of the peasants, even threw rocks at the Hanyuan county government offices. By now, cadres from all over Hanyuan and even from the larger municipality of Ya'an, more than four hours away by car, began to descend on the scene, as did hundreds of police reinforcements—some estimates put the number of People's Armed Police at ten thousand.[44] Up to this point, however, there had been no face-to-face meeting between the government officials and the protesters.[45] In the afternoon, the Hanyuan county magistrate, Zheng Shangkun, took the officials to the site. He told the protesters occupying the dam that he would work to resolve the issue and that the earthmoving would not start before then.[46] That night, the protesters went back to their villages, cautiously optimistic that their concerns would finally be met.[47] As in the past, they would soon be disappointed.

Local Government Reaction and Escalation of the Conflict

On October 30, 2004, the Hanyuan county government held a "five-level cadres" meeting (*wu ji ganbu hui*), including the vice secretary of the provincial Party committee, Li Chongxi, and the vice governor, Wang Huaichen, as well as cadres from the Sichuan, Ya'an, Hanyuan, and the *xiang* and *cun* (village) levels. Li and Wang gave speeches that differed substantially from the appeals that Zheng had made the day before. They proclaimed that the Pubugou project is one of the most important construction engineering

43. Ibid. The Associated Press reported, it turned out incorrectly, that the man had been beaten to death by the police. "One Killed in Mass Protest, October, 2004."

44. "Protesters Clash with Police in China over Land Requisitions: Up to 100,000 Farmers Clashed with Police in Southwest China, Protesting against Compensation Payments for Farmland Requisitioned to Make Way for the Pubugou Dam on the Dadu River in Sichuan," *Agence France-Presse*, November 1, 2004.

45. Protesters attempted to storm the Zhuangyuan Hotel, where central and provincial-level officials were staying. Epstein, "Proud of Protest."

46. Of course, he did not have the authority to say this. Some might say—indeed, my source strongly implied—that he was instructed to say this by the provincial-level officials so that the scene could be diffused without having to address the protesters' concerns.

47. Pubugou Interviewee, July 8, 2005.

projects in the country (*guojia de zhongdian gongcheng*). They further stated that the leaders of the opposition must voluntarily surrender to the public security bureau (PSB). Li and Wang also labeled the protests an "illegal act" (*feifa huodong*) and a case of "chaos" (*dongluan*), an extremely politically charged label, evoking the political discourse during the protests of 1989. There was such an environment of fear that some of the protest leaders did, in fact, voluntarily surrender, but the frustration would soon boil over again.

The first day of November saw the Hanyuan broadcasting station provide a summary announcement on the meeting. At the same time, between ten and twenty thousand police descended on the villages in the area and visited individual houses to warn people not to oppose the dam project. The Ya'an municipal government announced that the dam project would continue, a bulletin that was reprinted in all the local newspapers. The protesters were furious. That afternoon, they traveled to the dam site, walking, cycling, and using any transportation available to carry them the twenty kilometers. But this time the government was ready. They deployed a large number of People's Armed Police to create a human barrier to prevent access to the dam.[48] Over the course of the next five days, some forty to fifty thousand people descended on the area. This time, they were prepared for the cold weather, bringing tents and wearing winter clothing at night. The uneasy standoff continued.

On November 3, Sichuan Party Secretary Zhang Xuezhong, made another visit, coming straight from Chengdu to Dashu, calling a meeting to be attended by officials and representatives of the *yimin*. They catalogued the grievances of the *yimin* and acknowledged them to be legitimate. Despite this conciliatory gesture, after the meeting, Zhang attempted to leave the meeting hall but was prevented from doing so by a crowd of some thirty thousand people gathered outside. Police were eventually able to escort him out through the back where they all had to wade through the shallow frigid tributaries of the Dadu River. Zhang was reportedly furious.[49]

At around the same time, a crowd of ten thousand people was building temporary shelters at the dam site. On November 5, in the Hanyuan county seat, there were clashes between police and protesters, and reportedly a fair number of people were wounded, some seriously. There were also reports that a policeman was killed during skirmishes that had taken place the day before.[50] As the steady stream of news reports entreated the

48. The time before, the barrier was meant to prevent the protesters from going to the Heima headquarters of Guodian.
49. Pubugou Interviewee, July 8, 2005.
50. Ibid.

Figure 3.3 Protesters occupying the Pubugou Dam site, fall 2004. Photograph in author's possession.

protesters to return to their homes, armed police continued to be transferred to Hanyuan from other locales, some from as far away as neighboring Guizhou province.[51] The situation had spiraled to such a point that it required national-level intervention.

Dénouement

On November 6, the issue "went national" when CCP General Secretary Hu Jintao and Premier Wen Jiabao issued an internal report that contained their verdict on the events in Pubugou. Hu is reported to have said that although Pubugou is an important part of the "Develop the West" policy, the people's concerns must be adequately addressed and work on the dam could not begin until these issues were resolved. Wen went even further, echoing Hu's sentiments but adding that "the considerations of the peasants should be the first priority" (*kaolü nongmin de yiyi shi diyi*).

---

51. Pubugou Interviewee, July 10, 2005.

The situation seemed to quiet down the next day, November 7, when representatives from the Central Work Office, State Council Vice Secretary Wang Yang, and a vice minister from the Ministry of Public Security arrived in Hanyuan county. Because the protesters had by this time lost all confidence in the assurances of the local leaders, only reassurance from Beijing would suffice. After Wang Yang arrived, they went back to their homes.

The next day, there was a "six-level cadres" meeting (*liu ji ganbu hui*), which now included representatives from the Center. Wang Yang, representatives of the PSB *xitong*, and the Sichuan government all gave speeches. But the most important was that of Wang, who imparted the opinions of Hu Jintao and Wen Jiabao with the attendees. In conversations with others, Wang said that they would not permit "the investigation and punishment of those charged with guilt." This relaxed the atmosphere somewhat. Local people repeated their demand that the dam height be lowered. Wang said that he would look into this and have the technical experts make an assessment.[52]

The Political Lessons: November 2004–Present

Despite these assurances, it did not take long to reconsolidate control and to issue an official verdict on the protests. On the afternoon of November 9, the Liangshan prefecture Party committee secretary, Wu Jingping, convened an enlarged meeting of the standing committee, during which he disseminated documents from the Center (specifically, instructions issued by Politburo and State Council member and Minister of Public Security Zhou Yongkang) and to the provincial government (Deputy Sichuan Party Secretary Li Chongxi) on the handling of the protests. During this meeting, Wu emphasized first that problems should be "nipped in the bud" (*zhua miaotou*) through the establishment of an early warning mechanism that identifies such problems at their inception. He added that it was important to limit the damage of such an event through the establishment of better communication and reporting networks, including utilizing the Internet to report on unfolding events.

Finally, according to Wu, when such a mass event does occur, it is necessary to follow the principle of "one hand hard, one hand soft" (*yi shou ruan, yi shou ying*): "on the one hand, we must go down to the grass roots, to the masses, and win the majority's support by solving the specific problems raised by the masses, and handle the problems well. On the other hand, we must firmly

52. Pubugou Interviewee, July 8, 2005.

handle the situation according to the law as a powerful means, especially when the situation escalates to a certain degree." The basic theme was to maintain social stability and to identify the responsible parties quickly. Therefore, it would be necessary to use the grassroots Party infrastructure to resolve the problems at the village level (with bigger problems resolved at the township level), thus preventing problems from being transferred upwards.[53]

Playing the "good cop," the provincial authorities emphasized Hu Jintao's populist rhetoric. On November 14, an enlarged meeting of the provincial Party committee echoed several of the points of the earlier meeting, especially the importance of maintaining social stability and of preempting problems before they spiral out of control. The discussions in the meeting led to the resolution:

> Cadres must always feel deeply for the people while doing their tasks. They must seriously handle the people's lawful and reasonable requests. It is of crucial importance to stability that our cadres love the people, know their requests and carefully persuade the masses. Hu Jintao emphasized that no mass issue is small. Strong mass basis is the basis for having the ability to administrate, i.e., to win the people's heart. This requires our cadres to do mass persuasion work with feelings for them. We must be good at thinking in their shoes. We must do our best to handle their reasonable requests efficiently. As to those problems that we can't solve, we must patiently explain to them, and remove misunderstanding. Our cadres at each level will be able to do the work thoroughly and truly win the trust, understanding, and support of the people when they truly take the perspective of the people. Whatever we do, we must give adequate respect to and guarantee the people's interest, and if it is not beneficial to the vast majority or it is not supported or understood by the vast majority, then we must not do that thing.[54]

Starting in the middle of November, the Center and the Sichuan government started to conduct investigations of the local officials to see if there had been any wrongdoing. The Hanyuan Party secretary, Tan Zhengyu; two vice secretaries; the director of the Hanyuan Resettlement Office; and the mayor of Ya'an and former Hanyuan county Party secretary, Tang Fujing were all singled out for punishment and relieved of their duties. Bai Rangao, one the of the Hanyuan vice secretaries, was accused of embezzling 100,000 RMB, while Tang Fujing was accused of embezzling 4,800,000 RMB. This was welcome news to the local people, especially as Tang had

53. *Zhouwei juxing changwei kuangda huiyi bushu quanzhou weihu wending gongzuo* ["Prefecture Party Committee Convenes Large Standing Committee Meeting to Discuss Pacifying the Entire Prefecture and Stabilization Work"], *Liangshan Ribao*, November 10, 2004.

54. *Shengwei changwei kuangda hui qiangdiao* ["Provincial Party Committee Takes a Stand at Enlarged Standing Committee Meeting"], *Huaxibu Shibao*, November 14, 2004.

distinguished himself by saying that "whoever opposes the Pubugou project [should be aware that] the government opposes *you*" (*shei yao Pubugou guobuqu... zhengfu jiuhe ni guobuqu*).[55] On the other hand, they felt that the higher-level cadres were far more corrupt and had embezzled sums that dwarfed these amounts.

However, not all the punishment was reserved for dishonest officials. On November 28, there was a meeting to celebrate the protection of the Pubugou Heima construction camp. Attending the meeting were Guodian group Party committee member and vice general manager Chen Fei and other high-ranking Party members from the Guodian Dadu River Company and the Guodian Dadu River Company Pubugou subsidiary. The meeting featured another speech by Wu Jingping in which he pointed out that it was correct that the provincial Party committee and the provincial government defined the Pubugou event as an "organized, planned, and premeditated, large scale mass event." This event was not, he asserted, "purely a visit to the government driven by economic interest." Rather, it had become a "bad-natured political event and bad-natured social event... [which] directly endangered Ya'an city, Liangshan city, and even the entire province's social stability." Echoing previous comments, he said that "to protect Pubugou is to protect our own homes."

He went on to say that in handling this event, the Liangshan Party Committee and the government demonstrated that such protection of the construction site is a "test for the county Party committee and government's administrative ability, and the ability to face sudden events by the police force, armed police, and reserved force." This echoed Sichuan Party Secretary Zhang Xuezhong's call for "calm observation, calm response, focus on important points, flexible deployment of the forces, reduction of conflicts, and the solving of problems." By engaging in such a policy, it was possible to avoid bloodshed, to shield the masses from harm, to protect the police, and to hold onto the Heima construction camp unless it became absolutely necessary to retreat. Zhang argued that it was necessary to maintain the stability of the minority areas and to underscore the unity of the Yi minority with the Han. At the same time, it was necessary to demonstrate force in order to deter and to suppress "trouble makers."[56]

In April 2005, a vice director of the NDRC, Zhang Guobao, and a vice minister of the Land Resources Ministry, Lu Xinsi, went to Hanyuan for

55. Pubugou Interviewee, July 8, 2005.

56. *Wu Jingping qiangdiao: zengqiang zhizheng nengli baowei yifang ping'an* ["Wu Jingping Emphasizes Strengthening State Control and Safeguarding Peace"], *Liangshan Ribao*, November 9, 2004.

a meeting, as representatives of the Center. They said that it would be impossible to lower the dam height and disseminated propaganda materials to the local peasants explaining the policy. According to a knowledgeable source, when the villagers saw the cadres coming with these materials, they shut their doors in order to avoid receiving the documents. In addition, the authorities demanded that people sign a form certifying their acceptance of the dam. But they were met with stiff resistance.

In June, the Party secretary of Zhongba village (*cun*) of Dashu *zhen*, Wang Xiumin, was relieved of his duties, reportedly because he too had signaled his opposition to the way in which the *yimin* had been treated. This was meant as a warning to villagers against opposing the signing of these forms. However, as late as July 2005, many had refused to sign their names. Indeed, the political fallout led to a strengthening of governmental resolve to malign the protests, consolidate control, and see the project through. Subsequent reports from Hanyuan indicate that the entire area has since been papered over with propaganda posters reading, "The government is greatly concerned about reservoir area people," "The People's Liberation Army supports and protects the people," "Firm and unshakable support for building the Pubugou dam," and "Safe construction of the dam will help our civilization."[57]

A villager, Gao Qiansong, received a jail term of three years for his role in leading the protests, and a dozen others also received jail sentences. Up to four hundred others have been questioned by the local authorities.[58] In December 2006, Ran Tong, the lawyer assigned to Chen Tao, a man accused of killing a policeman during the protests, inquired about news of his client's appeal following his trial in April 2005. He was informed that his client had been executed.[59]

The Futility of Law

The shortcomings of official laws and regulations, taken alone, are illustrated by the compensation package that was ultimately adopted in the Pubugou case in May 2005. It was announced that a one-time moving expense (*banjia fei*) of 1,000 RMB per household for relocation within the county and 2,000 RMB for relocation outside the county (none were moved

57. Edward Lanfranco, "China's Hidden Hanyuan Incident," *United Press International*, November 25, 2004; and Luis Ramirez, "China's Sichuan Province Tense in Aftermath of Violent Anti-Dam Protests," *Voice of America*, November 24, 2004.

58. Shi Jiangtao, "Exodus Forced by Dam Under Way," *South China Morning Post*, December 13, 2005.

59. "China 'Executes Dam Protester,'" *BBC News*, December 7, 2006. Available at http://news.bbc.co.uk/2/hi/asia-pacific/6217148.stm, accessed May 19, 2007.

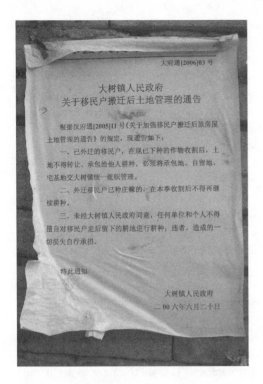

Figure 3.4 Post *yimin*-removal land use regulations, Dashu township, Hanyuan county. Photograph by the author, August 2006.

out of Sichuan province) would be paid out. Compensation per *mu* was arbitrarily established at 1,659 RMB by the Sichuan provincial government. There was no discussion with the *yimin;* this decision was made unilaterally by the Sichuan authorities. This 1,659 RMB was multiplied by 16 in a one-time payout, bringing the total payout per *mu* of 26,544 RMB.

This compensation package posed several problems. First, many considered the base estimate of 1,659 RMB to be too low. Second, the multiplier of 16 was far below the minimum standard of 30 as mandated by the December 3, 2004 "Guiding Opinion Regarding Perfecting a System for Levying of Resettlement Compensation" (*Guanyu wanshan zhengdi buchang anzhi zhidu de zhidao yijian*).[60] Unmoved, the Sichuan government said that the absolute highest compensation that it could afford was with the multiple of 16. Third, out of this 26,544 RMB, each individual only received 8,000 RMB. The remainder (more than 18,000 RMB per *mu*) was kept by the local authorities to purchase land and to build housing.

60. Pubugou Interviewee, October 31, 2005.

Not surprisingly, the local people were quite dissatisfied. Nevertheless, they were forced to sign letters expressing their acceptance of this agreement. Although through July 2005, many had not signed, the subsequent dismissal of Wang Xiumin, the one official who publicly signaled his opposition to the way in which the peasants were being compensated and forced to accept these terms, broke down this resistance. Since that time, many of those who had not signed have done so, largely out of fear.[61]

Further combinations of sticks and carrots have pushed local residents to accept their fate. Field trips to new homes seemed to reassure some of the *yimin*, although there are discrepancies between when they must vacate their current domiciles and when their new ones will be ready. They also face fines of up to 20,000 RMB if they do not leave their homes by the deadline.[62]

As of August 2006, the majority of those *yimin* to be moved out of Hanyuan altogether (27,900 out of a total of 33,000) have been moved out of the county to locales such as Mianyang, Wutongqiao, Leshan, and Ya'an. This has not been an easy process. For example, the *yimin* in Fulin township (*zhen cheng*) have been assigned to an area inexplicably slated to be submerged. The *yimin* have to now pay for things that were provided for free before, such as water and electricity, because of the higher altitude of the resettled location. And now there are plans to build a smaller dam that will further interfere with their access to water and resettlement more generally. Until these issues are ironed out, the people from Fulin are quite opposed to being resettled. However, if the people do not want to be resettled, they are confronted with PSB officials (*jincha*) to help "move them along." Every day, there are convoys of trucks to move the *yimin* property away from their homes and to their relocated sites. The police presence is based on a policy of "overwhelming force" (especially for those peasants being moved out of Hanyuan county) because they are afraid of a something as insignificant as a single piece of furniture falling from the trucks inciting a protest like in 2004. So, as the *yimin* are being moved and the moving trucks drive through Hanyuan city and such locales, there is often a phalanx of police on both sides of the road to ensure that any problems are preempted.[63]

Analysis of the Pubugou Case

Policy Entrepreneurs

With the possible exception of the protest "leaders," there were no policy entrepreneurs involved in the Pubugou case. As I have argued, these pro-

---

61. Ibid.
62. Shi, "Exodus Forced by Dam Under Way."
63. Pubugou Interviewee, August 9, 2006.

tests were marked by their spontaneity and their lack of organization beyond opposition to a single phenomenon, the dam at Pubugou. Moreover, beyond leadership in the sense of providing logistical aid to the protesters—temporary shelters, food, and so on—there does not seem to have been any attempt to engage the media or to coopt local leaders sympathetic to the protesters. In other words, policy entrepreneurs, so critical to the other two cases in this book, simply did not exist at Pubugou. Moreover, once the political verdicts on the Pubugou protests were leveled, it was ideologically forbidden for any aspiring policy entrepreneurs to remobilize the Hanyuan peasants.

Local government officials, though in many cases empathetic with the protesters, nonetheless were muzzled in terms of offering any meaningful support, whether overtly or through the leaking of documents or decisions. Indeed, the most brazen act on the part of sympathetic officials, one not to be minimized given the politically charged climate of 2004, was to try to get a more favorable compensation package. Any demands to lower the dam height were nonnegotiable. This might have been different if some more forward-looking policy entrepreneurs had pressed the claims of the *yimin* much earlier in the process. Moreover, the fate of Wang Xiumin certainly reinforced the notion among local officials that their inaction was probably a good thing for their political careers, if not for the actual *yimin*.

Conspicuous by their absence were nongovernmental organizations. Ironically, Hanyuan county is the location of DORS, an NGO committed to poverty alleviation. Apart from the fact that "poverty alleviation" was consistent with the rhetoric of the dam project (if not the reality) and was therefore not a wedge issue in any sense of the word, there was another dimension. DORS is expressly nonpolitical. They wisely kept themselves at some distance from the protests because failure to do so undoubtedly would have led to their closure by local authorities. (After initially welcoming me, DORS representatives politely but firmly refused to meet with me when I informed them of the nature of my research in the area.)

In general, NGOs played a very limited role. Individual NGO officers acted as intermediaries communicating what was unfolding on the ground to Chinese media outlets. Several of these individuals made visits to Hanyuan wearing their journalist hats in the summer of 2004. However, like the rest of the media, once events in Hanyuan became overtly political, these NGO representatives recognized their limited ability to influence them. The political "spaces" disappeared, replaced by the overwhelming imperative to maintain social stability.[64] By then, NGO officials, whose survival depends

64. Pubugou Interviewee, December 12, 2006.

on their political antennae, refused to come near the issue. Indeed, when I spoke to a particularly well-known and outspoken environmental NGO leader in 2006, even he refused to discuss Pubugou at all. One informed source vehemently denied (*kending meiyou*) any international influence on the Pubugou event.[65]

Without the necessary policy entrepreneurs in place and without NGO attention, it is not surprising that media attention would not be forthcoming, ensuring that Pubugou would become a watchword to be whispered among trusted friends and colleagues but unutterable in public. This had an enormous impact on the possibility of media framing for the opposition: it made any sort of unofficial framing impossible.

Issue Framing

The Pubugou story, at least as far as media coverage is concerned, can be broken down into two parts: before and after the protests. In the period between July and November 2004, stories emerged from the growing Hanyuan malaise. In July, at the same time that the *yimin* were petitioning the NDRC in Beijing, a group of NGO officers visited Hanyuan. On July 29, 2004, the *China Youth Daily* published a piece on the *yimin* that featured a section on Pubugou.[66] Another article that appeared in *Xinmin Zhoukan* about a week later featured a trip by NGO officers.[67] These articles initially helped increase the momentum for the protests in September 2004 because the demonstrators felt that their story was indeed getting out to a larger audience beyond Hanyuan.

The interplay between the media outlets and the protests entered a fascinating but extremely short-lived phase in late October. One month before, in September, a journalist named Tan Xinpeng arrived in Chengdu to write an article about Dujiangyan. He was told that there was an even more exciting story unfolding in real time in Hanyuan. Following this tip, Tan went to Hanyuan and wrote some pieces that appeared in the *China Youth Daily* on October 27, 2004.[68] The very next day, protesters occupied the dam site. To the delight of the crowd, one of the protesters read these articles to the other demonstrators through a megaphone. Certainly, it appears that this outside exposure to some degree emboldened the *yimin* and gave them

65. Ibid.

66. *Daxing shuidian gongcheng de "yingshang" burong hushi* ["Large-Scale Hydropower Engineering Project's Negative Effects Difficult to Overlook"]. Available at http://zqb.cyol.com/content/2004-07/29/content_918106.htm, accessed May 19, 2007.

67. *Xinmin Zhoukan*, August 2–8, 2004, 32–34.

68. Tan, *Dadu he shuidian yimin ju'e liyi liushui.*

(it turned out, false) comfort that their activities were being watched and favorably covered in the national media. Not long afterward, the *Ya'an Daily* rebutted Tan's *China Youth Daily* pieces.

A particularly brave journalist, Ouyang Bing, traveled to Hanyuan in the first days of November and wrote an article that came out in the *Fenghuang Zhoukan (Phoenix Weekly)* a few weeks later. After November 2004, there were no more articles that featured Pubugou beyond mentioning that protests had occurred a bit earlier (and which provided dubious claims, discussed below). Once the protests had taken place, the virtual absence of the media made it impossible to expand the policymaking sphere.[69]

This is due, in part, to the recasting of the Pubugou issue as a *political* problem (*zhengfu fangmian de wenti*). This official frame trumped any other possible framing of the issue, and the media took this as a cue to stay away. Although it would have made an extremely riveting story, the costs of reporting it were prohibitive. Not surprisingly, villagers were reluctant to speak to reporters, as suggested by the Associated Press:

> People who answered their phones at Hanyuan's hospital and a local bank said protests had occurred on Friday, but refused to give details, saying they'd been told by authorities not to talk to reporters. A woman who said she lived down the street from government offices also said a protest occurred, but wouldn't give details. A woman at the county government publicity office refused to answer questions, saying the official Xinhua News Agency would issue a report on the incident. None of the people contacted would give their names.[70]

The villagers' fears were not unfounded, as "a sea of green tents housing the thousands of paramilitary troops rushed in from the provincial capital" littered the area.[71] One soldier from Guizhou who had been dispatched to Pubugou said that there was a month-long telephone blackout, so he was unable to contact his family back home.[72] Toward the end of 2004, after the protests were over, National Public Radio reporter Rob Gifford was surrounded by plainclothes policemen and detained for several hours when he visited Hanyuan to report on Pubugou.[73]

69. Pubugou Interviewee, December 12, 2006.
70. "One Man Killed in Mass Protest."
71. Ramirez, "China's Sichuan Province Tense."
72. Mobilized soldiers are not allowed to have mobile telephones, so they had to rely on land lines, which were unavailable to them for the duration of the media blackout. Pubugou Interviewee, July 10, 2005.
73. Rob Gifford, "China Economic Policies Focus on Developing Rural Areas," National Public Radio, December 20, 2004. See www.npr.org/templates/story/story.php?storyId=4235919, accessed July 27, 2007.

In fact, what had occurred was the transition from one official state frame to another, from economic development to social stability. If the organizers of the protests at Pubugou might have had a chance of chiseling away at the former frame, they were powerless in the face of the latter. Once the authorities realized that these protests could not be put down by the normal means of local intimidation and indifference—that is, by demonstrating that the protesters were not being taken seriously and that they should disband—the response of the state had to be ratcheted up a few more notches. This took the form of labeling the protests a threat to social stability. It doesn't take the skills of a Kremlinologist to recognize the significance of this frame. Once it was invoked, all bets were essentially off. The local protesters found themselves isolated. The media and potential NGO allies stayed away. And any sympathetic local officials "swallowed their bitterness" (*chi ku*) that they felt toward the injustice levied at the *yimin*.

This news blackout did not prevent some stories from leaking out. However, some of these stories were so exaggerated that they created a general sense of confusion and undermined the credibility of those who felt that an injustice had taken place at Pubugou. For example, one activist who was present at the November 6 rallies phoned Beijing to say that "seventeen farmers were killed and forty wounded."[74] While a farmer told Luis Ramirez of Voice of America that "maybe more than 10,000 died. It cannot be estimated."[75] In the absence of credible information, it was easy for the government to downplay the story or even to deny that anything happened in Hanyuan at all. Such attacks on the credibility of government opponents have become as effective a strategy as muzzling the media, if not more so.

Broad Support for Policy Change

While there certainly was widespread support for policy change within the Hanyuan area, there was little meaningful support for it outside the region. Without policy entrepreneurs and the media to frame the issue, it didn't just lack resonance among the larger population, it was largely invisible. Indeed, it was more (or less) than that: it was politically taboo. Much like the protests in 1989, the authorities saw the Pubugou protests as an escalation of perhaps reasonable demands to a plane that simply could not be allowed to go further. As a result, Pubugou became a "nonevent."

In short, in the absence of policy entrepreneurs, there was no media frame for Pubugou other than the organic and largely self-evident frame of

74. "17 Farmers Shot Dead by Hanyuan Police," *Epoch Times*, November 12, 2004.
75. Ramirez, "China's Sichuan Province Tense."

social and economic injustice that might otherwise have been quite influential. Perhaps more important, potential policy entrepreneurs could not touch the issue because they understood the political significance of the "social stability" frame invoked by the government. As the next two cases make clear, policy entrepreneurs are most successful when their actions do not pose a threat to the state itself but rather seek to alter a particular policy. So policy entrepreneurs are wary of mobilizing people in a way that can be easily interpreted as a threat or that can scare away potential allies among the population more generally.

But there is another part to this story. Successful issue framing begets widespread support for policy change even as it taps into it. The government clampdown on media coverage is a necessary but insufficient explanation for the dearth of coverage on Pubugou. Another reason is that broad support for policy change in the Pubugou case has to do not with the government's alternative official frame of social stability, but with the traction of the unofficial frame of social injustice as expressed in the plight of the *yimin*. Unlike the Dujiangyan case described in the next chapter, which evoked cultural heritage frames, or the Nu River case, which was framed in terms of the environment, Pubugou's frame resonance was comparatively weak. Many people in China confront issues of social justice every day in their own lives. It is hard to extend this empathy to others, even those in far more difficult circumstances. Once the issue becomes economic, it seems to lose some of its luster, its power to attract the broad support necessary for policy change.[76]

In conclusion, Pubugou can—and must—be regarded as a failure of the protesters to establish and utilize the framework described in chapter 1. However, the opposition movement at Dujiangyan found almost complete success in exploiting this same framework. This is the subject of the next chapter.

76. Pubugou Interviewee, December 12, 2006.

4 | From Economic Development to Cultural Heritage: Expanding the Sphere at Dujiangyan

Should we sacrifice the heritage of the people and the world to the interests of some [*government*] departments?
—Representative of the Dujiangyan Cultural Relics Bureau

Although Dujiangyan is in Sichuan, the same province as Pubugou, with many of the same political actors (at least at the provincial level), and the events described in this chapter occurred barely a year before those of Pubugou, the political and policy outcomes could not have been more different. Indeed, the Dujiangyan case lies at the opposite endpoint to that of Pubugou: it marks the successful reversal of policy as a direct result of bottom-up opposition. Despite—or, more likely because of—the lack of protesters filling the streets, the Dujiangyan controversy was not prematurely snuffed out, and opponents were able to frame the issue in ways that proved to be impossible for the project proponents (or, rather, their political sponsors) to resist. In a sense, both Dujiangyan and Pubugou provide precedent-setting lessons to both sides of the debate. For dam supporters, they underscore the importance of preempting such opposition (such as the Yunnan government's insistence that the World Heritage designation for the Three Parallel Rivers region begin at an altitude of two thousand meters, thereby avoiding the same outcome as at Dujiangyan). For dam opponents, they highlight the importance of framing the issue in a way that tempers the nature of opposition so that it does not spin out of control and invite a government crackdown.

To understand the Dujiangyan case, one must first of all appreciate the resonance of Dujiangyan in the Chinese psyche. Dujiangyan (formerly *Guan*, literally "irrigation") county is the site of one of the world's most extraordinary premodern marvels of construction, the Dujiangyan Irrigation System. Conceived and constructed more than 2,250 years ago on the eve of the Qin Dynasty, the first to unify China, it represents an unequaled combination of technological expertise, environmental sensitivity, and engineering prowess, so much so that it serves the same functions today as it has since its inception. That it was completed only a few decades before the start of such construction projects as the Great Wall and the terracotta warriors in Xian ensures that Dujiangyan carries particular cultural weight with the Chinese.

The Dujiangyan Irrigation System

The Dujiangyan Irrigation System was completed in 251 B.C. It was a response to the unpredictability of the Min River, which had a tendency to overflow and flood Sichuan's otherwise fertile agricultural basin. In 276 B.C., Li Bin, the governor of Sichuan province, initiated the project, which would provide flood control and irrigation for the Chengdu Plain. The project itself can best be described as an attempt not to change the natural contours of the region's topography but rather to enhance them. The project made use of the natural characteristics of the Min River at Dujiangyan.

Figure 4.1 Dujiangyan Irrigation System. Photograph by the author, August 2004.

Despite this rich history and cultural symbolism, after 1949 the site was never safe from the grandiose plans of the Chinese leadership. In the 1950s, during a trip to Chengdu, Mao Zedong visited the Dujiangyan Irrigation System and expressed a desire to swim in the Min River.[1] Given the turbulence of the Min, Mao's handlers convinced the Chairman with some difficulty to give up his wish. Smarting from the "embarrassment" of this incident, Sichuan Party Secretary Li Jingquan, an ardent leftist supporter of Mao, ordered local officials to build a reservoir at the Yuzui ("fish mouth") section of Dujiangyan so that Mao could swim there on his next visit. Reportedly, Li pushed hard for the dam to be built, while the actual engineers for the project dragged their feet as much as they could, hoping that the order would be rescinded. They were concerned that their dam would obviate the functioning of the Dujiangyan Irrigation System.

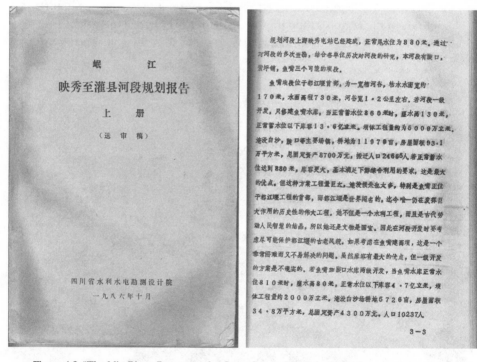

Figure 4.2 "The Min River, Report on the Guan County Watershed Plan."

1. Mao's swimming is the stuff of legend. Apart from building a swimming pool in his leadership compound in Zhongnanhai, thereafter known as the "swimming pool house" (*youyongchi*), where Mao entertained guests—reportedly including, on one unfortunate occasion, Nikita

According to an internal 1986 report, in 1955 and 1956, a plan was made for an eight-step development project along the upper reaches of the Min, of which the Zipingpu and Yuzui (later, Yangliuhu) dams were recommended as the first to be built. In 1958, construction began on the Yuzui dam, just four hundred meters above the Dujiangyan watershed, almost exactly the same spot as Yangliuhu. The Yuzui Dam would be part of the extended Zipingpu project to assist in flood control as well as to generate hydropower. In addition to providing Mao with a place to swim, the goal of the Zipingpu-Yuzui project was to provide flood control at the upper reaches of Min River, because when water exceeds a certain level, flooding can cause problems in the four main irrigation ditches connected with Dujiangyan. Started the same year as the Great Leap Forward, construction of the dams stopped in 1961. These plans were briefly resurrected in 1986, with Zipingpu and Yangliuhu referred to as "sister dams." But these plans never materialized—that is, until the early 2000s.[2]

For the engineers, much of the emphasis shifted to a gargantuan dam project at Zipingpu, seven kilometers upriver from Dujiangyan. In 2000, the State Environmental Protection Administration held a set of hearings on the Zipingpu proposal. These were attended by officials from the Ministry of Construction, the State Cultural Relics Bureau, the Ministry of Water Resources, and experts from the Chinese Academy of Social Sciences. The overall opinion of the Dujiangyan municipal government and its World Heritage Office, two key opponents to the Yangliuhu project, was mildly supportive of Zipingpu, or, at least, did not overtly or publicly oppose the project.[3] In subsequent hearings, experts who opposed the Zipingpu project were not invited; in any case, the press and NGOs were barred from all of these hearings. When challenged about the negative impact on the Dujiangyan Irrigation System an official reportedly said, "We'd rather lose the World Heritage status than the dam." The final decision was that Zipingpu would move ahead, while Yuzui would be tabled.[4] In February 2002, once the decision had been made and dam construction on Zipingpu had begun, Fauna and Flora International, an NGO, held a conference in Dujiangyan,

---

Khrushchev—Mao used his swimming to demonstrate his strength and vitality, as in his famous swim in the Yangtze, which provided one of the opening chess moves of the Cultural Revolution in 1966. Mao's swimming was a perennial headache to those charged with protecting the Chairman's safety. See Li Zhisui, *The Private Life of Chairman Mao*, New York: Random House, 1996.

2. "*Minjiang ying xiu zhi Guan xian heduan guihua baogao*" [The Min River, Report on the Guan County Watershed Plan], document in author's possession.

3. One source did say that some sort of "gag order" did exist in order to mute opposing views to the Zipingpu project.

4. Interview 06BJ02, March 10, 2006.

attended by more than three hundred people, in order to discuss how such heritage sites in China could be protected in the future.[5]

In May 2001, Chinese engineers had begun building the Zipingpu Dam. Zipingpu is one of China's "Ten Key Projects" currently under way; it was designed to increase hydropower and provide water to Chengdu, the provincial capital and to assist in the economic development of the western part of China more generally. A massive undertaking (the dam's height is estimated to be 156 meters) with a reported investment of 6.2 billion RMB, the Zipingpu project has been engineered to meet China's triple goals for any large dam-building project: hydropower, irrigation, and flood control. With regard to electricity generation, it is estimated to provide 550 million cubic meters of water to the Dujiangyan area and generate 3.4 billion kilowatt-hours of power annually. The reservoir itself will be able to hold 1.1 billion cubic meters of water, which would not only increase the acreage under irrigation from 672,670 to one million hectares, it would also increase the water supply to 50 cubic meters per second (28 during the dry season), easing the growing water shortages facing the provincial capital of Chengdu. Finally, Zipingpu is estimated to protect 720,000 people and 40,000 hectares of farmland from flooding. Construction has been moving apace, even exceeding its timetable.[6]

From "Yuzui" To "Yangliuhu"

However, by early 2003 the hydropower plans had become more complicated. The first indication came when the earth that had been removed from the Zipingpu began to accumulate a stone's throw from the Yuzui section of the Dujiangyan Irrigation System. Although the actual reason for this is shrouded in secrecy, many local officials opined that this signaled the covert construction of a hydropower station at Yangliuhu as part of the larger Zipingpu project (the renaming of the Yuzui project, which had been contemplated in the 1950s and the 1980s, as "Yangliuhu," fooled no one). This apparent secrecy supports the notion that Yangliuhu proponents anticipated a significant battle with those less supportive of the project and decided to preempt such a debate and present the Dujiangyan government with a fait accompli. The Yangliuhu Dam would be 1,200 meters wide and 23 meters high and be located 1,300 meters from the Yuzui

5. Interview 04BJ02, August 2, 2004.

6. "Damming of Yangtze River's Tributary Successful," *People's Daily Online*, November 24, 2002. Available at english.people.com.cn/200211/24/eng20021124_107373.shtml, accessed May 22, 2007.

section of Dujiangyan, with its waters coming as close as 350 meters to the Yuzui section of the Dujiangyan Irrigation System. With an investment of 500 million RMB, the Yangliuhu project was touted as "a necessary part of the Zipingpu hydropower project" once plans became public and opposition began to mount.

It was at this early stage that the secretive plans for Yangliuhu began to unravel. According to local Dujiangyan officials, the transfer of silt from Zipingpu to Yangliuhu occurred before any public or even internal appraisal of the Yangliuhu project had been made known to them and constituted, in their opinion, an "illegal act" (*weifa xingwei*). Indeed, there had been some reports of damage to the ecosystem brought about by the Zipingpu project, and these Dujiangyan officials felt that these problems should be addressed before any reports on the Yangliuhu project were even considered.[7] By April 8, however, "all preparations were complete," and the decision was made at the provincial level to move forward with the project.[8]

Certainly, officials in the Sichuan Cultural Relics Bureau (*Sichuan wenwu guanli ju*) were kept in the dark, as they never received a report soliciting their approval. An official told me that once he got wind of these plans, he went straight to the Sichuan Water Resources Bureau to discuss the matter. He was informed that Yangliuhu "absolutely had to be built" because it was necessary to raise the water level to meet the demands of the growing population in the provincial capital. The discussion became quite heated. According to this official, it was only when UNESCO's Beijing office informed him of the details of the project that he was able to make his case to the Sichuan provincial government. A vice governor convened a meeting, which included the Sichuan Development and Reform Commission, the Water Resources Bureau, the Dujiangyan Management Bureau (DMB), the Cultural Relics Bureau (CRB), and the Culture Bureau (CB). After a fiery debate, the Water Resources Bureau presented three separate plans with the Yangliuhu dam 1, 1.5, and 2 kilometers, respectively, from Yuzui. The CRB and the BC rejected all three plans.[9]

As soon as details of the project started to leak beyond official channels, the response was swift, overwhelming, and negative. In part as a response to the emerging criticisms, on April 28 the DMB convened a meeting of engineers and other experts to discuss the Yangliuhu project. Attending the conference was a group of government representatives from the following units: the Dujiangyan World Heritage Office, the Dujiangyan EPA,

7. Interviews 04DJY02 and 04DJY03, August 6, 2004.
8. Interview 04DJY01, August 5, 2004.
9. Interview 05CD01, July 7, 2005.

and the CRB (other units, including the Seismological Bureau, were also in attendance), along with their provincial counterparts as well as representatives from the provincial bureaus of construction (*jianshe*) and planning (*guihua*). The reasons for the intense opposition to Yangliuhu became crystallized at this conference.

First, a dam and hydropower station at Yangliuhu would negatively impact the diverse ecosystem of the Dujiangyan Qingchengshan area. This would effectively destroy a World Heritage site with deep cultural connections to the Chinese psyche. Because of its proximity to the Yuzui portion of Dujiangyan, it would create "vision pollution" (*shijue wuran*) and would rob the Dujiangyan government of much-needed revenue (at more than 60 RMB, individual ticket prices for Dujiangyan are quite high). Dujiangyan does not have any industry of its own, and although unemployment statistics are unavailable, Dujiangyan officials conceded that there is significant unemployment. Yet the Dujiangyan government was recently upgraded in administrative rank, from a county (*xian*) to a county-level municipality (*xian ji shi*). This was a result of Dujiangyan's economic development, which in turn could only be a result of its tourism trade. And this would be curtailed if the Dujiangyan Irrigation System were made inoperable. So there was a significant financial stake in the continuation of this World Heritage site.

Second, because the Dujiangyan Irrigation System continues its long-standing functions of flood control and irrigation, Yangliuhu would negatively affect the entire Sichuan Basin and the agricultural output of Sichuan province more generally. Finally, Yangliuhu went against many organizational interests of Dujiangyan's bureaucratic units and appeared to violate the Environmental Protection and Cultural Relics laws as well as other regulations, including the "Sichuan Provincial Regulations for the Protection of World Heritage Sites" (*Sichuan sheng yichan baohu tiaoli*).

Opponents framed the issue shrewdly, anticipating the positions to be taken by supporters of the project as well as playing to the biases of undecided decisionmakers. One of the arguments underscored the folly of blindly following the West in wanton dam construction. Such a project, argued the opponents, would only last a couple of hundred years, after which silt accumulation would make the dam inoperable. Thus, argued these experts, this is really a question of short-term versus long-term interest. While compelling from a relatively objective point of view, this issue frame was not terribly persuasive politically. By this point, however, another, far more potent media frame was being constructed, having to do with cultural heritage issues. It ultimately carried the day.

Policy Entrepreneurs in Action

In June 2003, activist-journalists Zhang Kejia and Wang Yongchen were sponsored by Conservation International to visit Mugecuo Lake in western Sichuan. Earlier that year, local residents from the Mugecuo Lake area near Gonggashan Mountain wrote a letter to Premier Wen Jiabao, urging him to take into account their opinions that Mugecuo Lake should not be ruined in order to build a hydropower project. Wen's response was that the issue should be reviewed before going forward. However, the Ministry of Water Resources and, surprisingly, SEPA and the Tourism Bureau all agreed that the project should move forward.[10]

When Zhang and Wang went to Mugecuo in June 2003, the Kangding prefecture government ordered the villagers not to say anything negative to the "outsiders." As they were leaving, however, one villager ran after them, telling them that the government line was inaccurate and that the villagers all opposed the project. With this incident in mind, Zhang and Wang walked right into the Yangliuhu controversy. On their way back from Kangding to Beijing, they stopped in Chengdu, where an official from the Dujiangyan World Heritage Office asked them to come to Dujiangyan so that they could see what was going on with the Yangliuhu project and the controversy surrounding it.[11]

On June 5, at another conference sponsored by proponents of the Yangliuhu project, it was announced that the project would move forward. Local experts and the Dujiangyan World Heritage Office, however, disagreed with this assessment. Shrewdly, the Dujiangyan World Heritage Office appealed to the Sichuan Construction Bureau, which sided with the opponents of the project.[12] But some units at the provincial level were still leaning in favor of the project. The two journalists were in attendance. While in Dujiangyan, they were pulled aside by municipal officials and told that this was a big story and that only somebody in their

10. In November 2006 it was announced that the project would be scrapped. See International Rivers Network, "Chinese Prefecture Cancels Dam Project on Sacred Tibetan Lake," press release, available at www.irn.org/programs/china/index.php?id=archive/061113cancel. html, accessed May 22, 2007.

11. Interview 06BJ02, March 10, 2006.

12. The construction bureaucracy is a relatively new one and is preoccupied with increasing its administrative power, much like the Quality Technical Supervision Bureaucracy. See Mertha, *Politics of Piracy;* and Interview 04BJ01, August 2, 2004. As such, it strategically picks its battles on those issues that provide something to gain politically as well as substantively. Politically, being on the winning side of a coalition brings certain benefits; substantively, reservoirs do tend to flood areas in which roads and other infrastructure projects could be built. The construction bureaucracy does not have a hand in dam construction. See also Joseph Kahn, "Can China Reform Itself?" *New York Times,* July 8, 2007.

position as journalists with their ability to provide national-level exposure would be able to initiate the necessary momentum; they were told to contact the UNESCO offices in Beijing for further information. Upon returning to Beijing, Zhang and Wang quickly discovered UNESCO's reluctance to "rock the boat."[13]

While they were in Dujiangyan, however, the Dujiangyan World Heritage Office laid out the full panoply of issues at stake. Other Dujiangyan officials who were opposed to the project carefully leaked additional information. The article itself, "World Heritage Site in Dujiangyan, New Dam Construction There Can Be No Turning Back" (*shijie wenhua yichan Dujiangyan zaijian xinba jianzai xianshang*), was published in the *China Youth Daily* on July 7, 2003. The article had the effect of framing the issue not in terms of economic development (always an enticing line of argument) but rather as an attack on China's own cultural heritage. Moreover, it simultaneously played to a national and international audience, particularly because *China Youth Daily* articles are simultaneously posted on the Internet.

This opened the floodgates, as other actors got involved in the process at the local and national levels. On June 27, 2003, the Dujiangyan Cultural Relics Bureau, together with the provincial government, issued a report, "Bulletin Regarding the Situation of the Dujiangyan Yangliuhu Hydropower Construction for the Beijing Office of UNESCO" (*guanyu lianheguo jiaokewen zuzhi zhu Beijing banshichu, guanzhu Dujiangyan Yangliuhu shuili gongcheng qingkuang de baogao*). Less than three weeks later, the Standing Committee of the Dujiangyan Municipal Party Committee signaled its support.[14]

Local Dujiangyan officials continued to bring the issue to national attention and to mobilize the press. Throughout July and August, interviews were granted to the Guangdong-based *Southern Weekend* (*Nanfang Zhoumou*), a paper noted for its zeal in airing the government's dirty laundry, as well as other media outlets throughout the country, including Shanghai, Hebei, and, of course, Beijing. From July through September, a single official in Dujiangyan had been interviewed by "more than one hundred" newspapers and Internet outlets. This official cleverly couched his opposition in cultural heritage terms, asking his interviewers rhetorically, "Should we sacrifice the heritage of the people and the world to the interests of

---

13. Interview 06BJ02, March 10, 2006.

14. "*Baohu shijie wenhua yichan, fandui xiujian Yangliuhu shuiku*" [Protect World Heritage Sites, Oppose the Construction of the Yangliuhu Reservoir], internal document, Dujiangyan Municipal Cultural Relics Bureau, December 22, 2003.

some [political] departments?" These reports quickly found their way to the Internet.[15]

Authorities in Beijing and in Chengdu were paying close attention to these developments; indeed, given the momentum they generated and the opposition that arose, they could ill afford not to. On August 26, the provincial DPC submitted a report to the provincial governor. The recommendations of the report were that the project should be examined carefully and that the plan should be suspended until a consensus could be reached. Three days later, these recommendations were supplanted by an order of the governor to the effect that the project be abandoned.[16] According to several experts, this was the first time in the history of the People's Republic of China that a decision on an engineering project of such magnitude—a decision that had already been reached—was reversed. Just as significantly, experts and the media formed the critical core of the opposition in terms of the larger contours of the debate. This was very much in contrast to the dam-building projects of the 1980s in China, to the Pubugou case, and, indeed, to much of the political process as we understand it in China today.

15. Interview 04DJY02, August 6, 2004. Here is a partial list of articles: *Poyu gefang yali Dujiangyan guanli ju jueding zanting Yangliuhu gongcheng* ["Constrained on All Sides, the Dujiangyan Management Bureau Decides to Suspend Temporarily the Yangliuhu Engineering Project"], *Zhongguo xinwen wang*, August 7, 2003, available at www.chinanews.com.cn/n/2003-08-07/26/333016.html; *Zhuanjia xuezhe he minzhong dui Dujiangyan jianba shuo bu zhenglun ruhuo rutu* ["Experts, Scholars, and Masses Say the Dujiangyan Dam Construction Dispute Should Not Be Fiery or Bitter"], *Zhongguo xinwen wang*, August 7, 2003, available at www.chinanews.com.cn/n/2003-08-07/26/332883.html; *Dujiangyan ni zai jian xin ba shou zhiyi zhuanjia diaocha zu ji qianwang diaocha* ["Draft Plan for New Dujiangyan Dam Called into Question, Expert Investigation Unit to Examine"], *Zhongguo xinwen wang*, August 3, 2003, available at www.chinanews.com.cn/n/2003-08-03/26/331147.html; *Shijie yichan Dujiangyan yao jian xin ba shou piping, lianheguo guanyuan guanzhu* ["Dam at Dujiangyan World Heritage Site Draws Criticism, UN Officials Follow the Story"], *Zhongguo xinwen wang*, July 9, 2003, available at www.chinanews.com.cn/n/2003-07-09/26/322426.html; and *Buneng digu Yangliuhu gongcheng dui Dujiangyan de yingxiang* ["The Influence of the Yangliuhu Engineering Project on Dujiangyan Must Not Be Underestimated"], *Sichuan zaixian*, August 5, 2003, available at sichuan.scol.com.cn/bsxw/20030805/20038551245_hx.htm. On the pro-Yangliuhu side, see *Shuili bu zhuanjia Jing Renyu: Dujiangyan Yangliuhu gongcheng yiding hui shangma* ["Ministry of Water Resources Expert Jing Renyu: Dujiangyan Yangliuhu Engineering Project Must Commence"], *Henan baoye*, August 9, 2003, available at news.tom.com/Archive/1002/2003/8/9-26366.html; and Li Yingfa, *Zunzhong lishi miandui xianshi—xiujian Yangliuhu dui Dujiangyan wenhua yichan yingxiang de duice* ["Cherishing History Confronts Reality: The Countermeasures against the Yangliuhu Construction's Effect on Dujiangyan Cultural Heritage"], *Sichuan shuili*, June 2003, available at scholar.ilib.cn/abstract.aspx?A=scsl200306003. All Internet articles accessed May 22, 2007.

16. Interview 06BJ02, March 10, 2006. It should be noted that the Sichuan government appeared to take the lead in the Dujiangyan case, while the Sichuan Party Committee was at the forefront of the Pubugou case discussed in chapter 3. This is in no small part because Pubugou quickly become political as a result of the widespread demonstrations, causing it to

Analysis of the Dujiangyan Case

We begin with the bifurcation of the official and unofficial views of the various individuals in government directly involved in the controversy. In a very different context, Perry Link has written about the typical practice in China whereby divergent public and private points of view among officials are kept internal.[17] Others have also underscored the "irresistibility" of official mandates even when they come up against the genuine preferences of the officials charged with pursuing such mandates.[18] As Dai Qing has amply illustrated, the list of opponents to the Three Gorges Dam Project was long and distinguished, yet these individuals were unable to get their views into the broader debate; in some cases, they were actually punished for espousing such views even within the limited scope that they were able to do so.[19] In the Dujiangyan case, by contrast, local and national officials, understanding that their sanctioned positions prevented them from articulating personal points of view or even objective facts, communicated—leaked—this information to the press in order to break the monopoly on information held by the government agencies that supported the project.

Second, apart from the First Five Year Plan (1953–1957), politics in China after 1949 can be seen as a series of experiments in decentralization: top leaders sensing opposition from their colleagues bypassing these impediments by appealing to an increasingly broad coalition of decentralized actors. During the Great Leap Forward (1958–1961), Mao Zedong leapfrogged over the central ministries by appealing directly to local and grassroots Party activists. In the mid- to late 1960s, he went one better, circumventing the already widely decentralized Party apparatus by appealing directly to China's youth, thus mobilizing what became the Red Guard movement during the radical phase of the Cultural Revolution (1966–1969).[20] This was not limited to the Mao era, however. As Susan Shirk has pointed out, Deng Xiaoping was only able to push through his reform program in 1979 by "playing to the provinces"—bypassing truculent officials in Beijing by packing the Central Committee with local actors, a tactic also

require a response from the Party apparatus, not simply government organs. In Dujiangyan, debate remained largely within the bounds of acceptable political behavior, so the Party did not involve itself as much. Pubugou Interviewee, December 12, 2006.

17. Perry Link, *Evening Chats in Beijing: Probing China's Predicament,* New York: Norton, 1992, esp. ch. 4.

18. Graham Allison and Philip Zelikow, *Essence of Decision: Explaining the Cuban Missile Crisis,* 2nd ed., New York: Longman, 1999, 307.

19. Dai, ed., *Yangtze! Yangtze!*

20. The irony, of course, is that in order to quell Red Guard violence, Mao had to bring in the military to restore order, effectively recentralizing the state to some degree.

used by Mao.[21] In the case of Yangliuhu, this dynamic is the same, except with one significant difference: the *direction* of such appeals. This was not a "top-down" dynamic but rather a "bottom-out" one. Put simply, status quo powers—those in favor of the Yangliuhu project—at the middle—provincial and ministry (*bu ji*)—administrative levels were surrounded by a decentralized opposition made possible by taking information out of the hands of those who had so jealously guarded it and making it available to—potentially—anybody who could get ahold of a newspaper or surf the Internet.

Policy Entrepreneurs

The core of the opposition at the municipal level included the Dujiangyan CRB, the Dujiangyan World Heritage Office, the Dujiangyan EPA, and the Dujiangyan Seismological Bureau (because of the instability that reservoirs incur over seismological fault lines). In addition, the project was opposed by the generalists in the Dujiangyan municipal government, which has principal leadership relations over these bureaus.

Although one can only speculate about the preferences and motivations of these critical actors, some goals and predispositions are more readily apparent. Among the principal movers and shakers were officials within the Dujiangyan CRB. It is unclear what role personal ambitions played in the process, but it seems that many of these officials' actions can be explained by the organizational goals of the office in which they worked. Yangliuhu provided an opportunity for this bureaucracy to maintain or even increase its strength. These officials could argue that their strong opposition to the project was based on the organizational goals of the office. Indeed, Bian Zaibin would have been remiss in his duties as director of the CRB if he had not pursued this with the requisite zeal and vigor. Moreover, given that his immediate superior in this regard was the Dujiangyan municipal government—which was also opposed to the project—Bian could rely on some degree of support from his superiors.

The case of journalists like Zhang Kejia is somewhat more complicated. On the one hand, she is an editor of a government mouthpiece, the *China Youth Daily*. On the other hand, her sober pro-environmental positions are well-known and largely unassailable. Her dual roles as journalist and activist allow her to use her newspaper as a "bully pulpit" to disseminate her concerns to the wider public. But much of her success in the Dujiangyan case had to do with the choice of frame—not the environment but rather the potential loss of China's cultural heritage.

21. Shirk, *The Political Logic of Economic Reform in China.*

Another key player was the Dujiangyan World Heritage Office, a unit directly under the control of the Dujiangyan government. At first glance, the director of the office, Deng Chongzhu, does not seem all that powerful. His placement at the head of the Dujiangyan World Heritage Office was a "golden handshake" for retired cadres. However, Deng's previous positions included secretary of the Dujiangyan Party committee (*Dujiangyan shiwei shuji*), the most powerful posting in Dujiangyan. This position, during which Deng cultivated relationships with a wide array of officials inside and outside Dujiangyan, seems to have given him a considerable degree of insulation and unofficial power in pursuing his goal of preserving Dujiangyan. Deng's case may well provide a cautionary note to those who believe that providing retired cadres with largely "symbolic" or "ceremonial" portfolios is an effective way of "getting them out of politics."

Issue Framing

Obviously, the media was absolutely critical to the success of the anti-Yangliuhu coalition. Around 180 media outlets, newspapers, magazines, and TV and radio stations reported on the controversy, and almost all of them sided with the opposition.[22] Dujiangyan officials often provided the information to the media, but the media itself made the decision to actually run with the story. As noted above, there are several facets to understanding how this occurred.

The first is the exponential growth of media outlets in China. There are now thousands of newspapers and periodicals. This has been accompanied by the "marketization" of such previously labeled "thought work."[23] Rather, they must appeal to Chinese consumers, who, like people anywhere, tend to prefer racier stuff to government boilerplate. Of course, such progress in not exactly linear. Jiang Yanyong, the doctor who helped pry open the press during the SARS crisis in 2003, was placed under a form of house arrest during the fifteenth anniversary of the Tiananmen crackdown. Nevertheless, it has become increasingly difficult for the authorities to control the press, especially with a story as "juicy" as this one that appeals to precisely the type of nationalism kindled by the current leadership in the absence of coherent socialist ideological appeals.

Not only is Dujiangyan a World Heritage site, which makes it a source of pride for China and the Chinese; this designation is an independent (external, international) acknowledgment that Dujiangyan is, in fact, a

22. Kelly Haggert and Mu Lan, "People Power Sinks a Dam," *Three Gorges Probe*, October 16, 2003.
23. Daniel C. Lynch, *After the Propaganda State: Media, Politics, and "Thought Work" in Reformed China*, Stanford, CA: Stanford University Press, 1999.

central, symbolic part of the cultural heritage of what makes China China, what makes the Chinese Chinese. It was only by mobilizing these elements of Dujiangyan that the opposition was able to "interrupt" the seemingly otherwise immovable "equilibrium" of western development in China. As such, it was a brilliant political strategy on the part of the opposition.[24] Few issues can unseat a mantra as powerful to contemporary China as that of "economic development"; "China's cultural heritage" is one of those that can.

Aside from its symbolic significance and resonance, this frame has some other qualities. Protecting China's cultural heritage does not conflict with the broad policy lines of the state or the party, even if it does frustrate some of these institutions' short-term goals. Indeed, it is a clever issue because it effortlessly ties into larger notions of nationalism, the one "-ism" that Beijing has allowed to coexist alongside Marxism, communism, and socialism in the official political discourse.[25]

Moreover, even though the Yangliuhu issue resonated in a very deep and diffuse way, the policy issue itself was fairly narrow and relatively easy to "fix." The sunk costs for Yangliuhu were quite low, making it easier for the government to make a concession. As the state was not the target, it was easy to dispose of the policy once it became so unpopular within mainstream Chinese society.

Finally, the issue "went national" extraordinarily quickly, thanks to an aggressive journalism corps as well as local policy entrepreneurs in Dujiangyan. By doing so, the issue reached a large audience, both domestically and internationally, and could not be easily quashed. This ensured that this issue would remain "in play" until the government either dug in its heels and moved forward, risking widespread protest and condemnation within China, or gave in to the anti-Yangliuhu camp. The government decided to give in—in the appropriate aquatic terminology, "to reverse course."

Broad Support for Policy Change

Finally, the public at large played an important role as "consumers" of the debate. Because Dujiangyan plays such a prominent role in the Chinese psyche, the public was easily mobilizable—or at least the ultimate decisionmakers anticipated that it would be so. As such, the public would have been

---

24. Not only that: it also expertly exploited the Chinese government's more general preoccupation with cultural nationalism as a way to fill the void left by the ideological bankruptcy of Marxism, Leninism, and Mao Zedong Thought.

25. Peter Hays Gries, *China's New Nationalism: Pride, Politics, and Diplomacy*, Berkeley: University of California Press, 2004.

extremely difficult to placate if the decision to proceed at Yangliuhu had gone forward.

In the absence of reliable survey data, one must rely on educated inferences based on careful study of the situation. Obviously, China's leaders are not beholden to regular elections beyond the "selectorate" of Party and national people's congresses; nevertheless, in domestic matters as well as international ones, China's leadership's primary constituency is its domestic one. That is to say, the leadership in Zhongnanhai pays attention to reports of "public opinion," as measured indirectly by what the media produces—a premise that assumes the media to be increasingly consumer-driven, which is the case. Thus, an important proxy measure of public opinion in China is the amount of media coverage on a given issue.[26] Given that, as noted above, somewhere in the neighborhood of 180 media outlets reported on the controversy, almost all of them siding with the opposition, the message to the top leadership was clear.

In addition, the Center is acutely aware of problems in managing the state and relies on several channels of information to assist it in governing more effectively and in making decisions that might not be politically viable. The Chinese government relies in part on information from outside its immediate purview to rein in corrupt officials, to uncover projects that may incur costs hidden from official assessments, and to monitor the pulse of the Chinese public. This is where "public opinion," insofar as it can be identified and categorized as such, comes into play.

The Dujiangyan case was quickly swept up by the media and became such a cause célèbre that according to one well-placed source within the State Council apparatus, Premier Wen Jiabao commented in March of 2005 that "for a project *which has aroused such public concern*, we need to devote more time and make assessments based on scientific considerations."[27] So while widespread support for policy change is not a substitute for top-level decisionmaking, it is not inimical to it. Indeed, it is better to see the two as an increasing dynamic of interaction within the Chinese policy process.

By way of conclusion, it is important to neutralize a potential alternative explanation: activism solely on the part of the government, which would render the nongovernmental players in the above account largely irrelevant. Specifically, the success of the anti-Yangliuhu coalition stands in sharp contrast to the failure to address the concerns of the people relocated as a result of the Zipingpu project, which preceded the Yangliuhu controversy.

26. Interview 05BJ04, July 6, 2005.
27. Ibid.

For instance, one can argue that the actors within the Dujiangyan government are by nature "enlightened" and thus would take on other, related issues to Dujiangyan/Yangliuhu. But this alternative argument does not bear scrutiny. The plight of the *yimin* relocated from Zipingpu is a case in point. After the Yangliuhu controversy died down, Dujiangyan returned to business as usual. But the *yimin* issue that has arisen is notable for the absence of the type of sphere expansion that was perfected during the Yangliuhu debate. According to a local official, some of the *yimin* have already been resettled in Chongbai municipality, but the conditions of their new housing and attached land is unknown. More to the point, many of these resettled people have not yet been assigned housing and instead have been forced to move in with relatives or otherwise fend for themselves. Indeed, this resembles the situation that led to the protests in Pubugou much more than the political dynamics in Dujiangyan only a few years before.[28]

The cases in chapters 3 and 4 laid out the endpoints of the spectrum of policy outcomes vis-à-vis the framework developed in chapter 1. Although they are not unique cases, they are far from the norm. Anti-hydropower opposition is less likely to be put down in the manner illustrated by Pubugou simply because of the desire to prevent the mushrooming of such politically sensitive processes and outcomes. The situation in Dujiangyan is relatively rare, not necessarily with regard to the outcome but because of the rapid pace of the resolution of the conflict. A more drawn-out case would be more representative of the increasing give-and-take that is the nature of China's political pluralization within the politics of hydropower. Such a case, the Nu River Project, is the subject of the next chapter.

28. Pubugou Interviewee, October 31, 2005.

5 | The Nu River Project and the Middle Ground of Political Pluralization

This, then, may be the legacy of the ongoing protests against hydroelectric schemes in northwest Yunnan: the mixing of hopes and despair, triumphs and struggle.
—RALPH LITZINGER, in *Grassroots Political Reform in Contemporary China*, 2007

Pubugou and Dujiangyan provide the two extremes of the constraints and opportunities facing those who support the political pluralization of hydropower policy in China today. Oftentimes, policy change occurs along the margins of the policy process instead of at the core. Rather than indicate a relative lack of pluralization, what such incremental policy change may suggest is precisely the opposite: a growing complexity of the political processes of hydropower policy formation. Some elements of what occurred at Dujiangyan—and which were completely absent at Pubugou—are embedded in the policy process of the Nu River Project (NRP) in Yunnan province. In this chapter, I focus on the NRP to illustrate the middle ground, the gray area that exists between more ideal-type cases examined in the preceding two chapters.

Antecedent: The Manwan Dam on the Lancang River

The Nu River controversy is the latest in a growing series of hydropower projects, including those along the Lancang and Jinsha rivers, aimed at developing perennially underdeveloped Yunnan province. In the 1980s, much of Yunnan was crisscrossed with dirt roads, and its infrastructure was all too reminiscent of the days when "Vinegar Joe" Stilwell and Claire Chennault and his Flying Tigers walked the streets of Kunming in the 1940s. As then, Yunnan today experiences chronic and severe electricity shortages.

The earliest of the large-scale hydropower projects in Yunnan province, the Manwan Dam along the Lancang River, was begun in 1985 and finished in 1994. At the time, there was a significant constraint: no single government entity could provide enough investment to underwrite the entire project. Rather, the Yunnan provincial government entered into a joint venture with several state-owned enterprises housed under the Ministry of Energy. At the time, all the decision-making came from the national and provincial-level governments. There was no impact assessment analysis, since the Environmental Impact Assessment Law was two decades away from being promulgated; the government guaranteed a resettlement plan even though it knew it did not have the adequate funds (it promised to "do [its] best"). The ambiguous goal was that government investment would make the people in the region "better off." But for many, the outcome was exactly the opposite: the government generated considerable income from taxation, and the people fell deeper into poverty. And silting became a huge problem. After only one year of operation, research concluded that one year's siltation surpassed the anticipated projection by a factor of five.[1] Yet at the time, there was no outlet for opposition to the project. Twenty years later, however, the situation had changed dramatically.

The Manwan Dam and the "817 Incident"

The fortunes of the peasants (*yimin*) resettled as a result of the construction of the Manwan Dam are important because it sets the context of the debate over the Nu River in two ways. First, like the Three Gorges, the Manwan Dam provides a baseline for comparison with the Nu River controversy. Second, as the next section will show, the anger and resentment among those negatively affected by the handling of the Manwan Dam have bled into the present day and have contributed significantly to the momentum necessary for policy entrepreneurs to make their case against the Nu River Project (NRP).

Actual construction on the Manwan Dam started in 1987; it was operating in full by 1994. Agreements between the Manwan Power Station Management Office and the local governments notwithstanding, affordable electricity remained out of the hands of the local peasants. Moreover, promises of compensation for the *yimin* were not kept. Finally, the relatively poor planning of the dam resulted in a series of negative ecological effects, further diminishing the livelihood of local residents.[2] However, as was the case

1. Interview 05KM03A, April 28, 2005.
2. Magee, *New Energy Geographies.*

with Zipingpu or with Pubugou before 2004, there was little organized opposition. The Manwan Dam project encountered little resistance from the local people, the NGOs, the media, environmental bureaus, or other members of the anti-dam coalition described in previous chapters.

The Manwan issue reappeared on the political radar after the government announced a power station in Xiaowan, a similar dam project along the Lancang River. Unlike the case with the Manwan project, peasant activists were ready to mobilize, led in part by the NGO Green Watershed (GW) under the direction of Yu Xiaogang. On August 17, 2003—"817"—some three thousand people affected by the Manwan Dam sought a meeting with the Manwan Huaneng Power Company to air their grievances. How they came together remains somewhat unclear, but as an organized group they were able to mobilize many angry protesters without any damage, destruction, or violence. The demonstration lasted three days. The local Jingdong and Yuxian county and Manwan township governments promised to resolve the problems laid out by the protesters, mostly related to inadequate compensation for resettlement. Nevertheless, after the incident, there was no further progress on the issue. The government hemmed and hawed, and while it acknowledged that the protest was peaceful, it could not allow such a protest to recur.

But recur it did. On the first anniversary of the 817 Incident, a meeting was organized, in part by Green Watershed, to celebrate the 817 Incident and to draw lessons from it. There were 150 representatives of local communities in attendance. The local government got wind of this and "asked if they could attend." GW organizers sensed a trap and told the local participants to be vigilant and to make sure to abide by the laws lest they get in trouble on some trumped-up charges. GW representatives were told that if they pursued such an open, public meeting, they would have to secure government approval first—permission that might well be denied by the local government. However, if the meeting was not a "public" demonstration but a "private" meeting, they could bypass the requirement of securing government permission and meet without breaking the law.

They decided to hold the meeting in a restaurant, thereby turning it into a "private" function. But the restaurant in which they were to have the meeting did not have a large enough roof to accommodate so many people, creating the risk that the authorities would label the meeting a "public demonstration," so the organizers decided to build a temporary addition to the roof so that everybody could fit underneath. This came to the attention of the local Public Security Bureau (PSB), which informed GW that they were organizing an illegal meeting, but GW representatives told the PSB that they should check the laws governing public protests and calmly demonstrated that what they were doing was within the law. The PSB officials reluctantly allowed the meeting to continue.

To augment the participation of local people, GW underwrote the transportation costs for people from other areas affected by the Manwan and other dams, including those threatened by the NRP. Six people came from the Nujiang area, and another six came from the Jinsha River area. Yu Xiaogang had emphasized to the participants that the meeting should be understood as a "workshop" to present and discuss the issues affecting the people from the Lancang, Jinsha, and Nu rivers who were likely to be resettled and to find ways to resolve them. It was not to be considered a platform from which to launch a demonstration, like the one at Pubugou.

During the meeting for the first anniversary of 817, people broke off into discussion groups to debate the issues facing them, and each group compiled a report. For example, one such issue had to do with an article that had appeared in a single newspaper (in Yunnan's provincial capital Kunming, *not* in the local affected areas) that detailed how local Manwan government corruption had funneled five million RMB of resettlement money to build a hotel and to purchase luxury cars. GW provided copies of the article to the people in the Manwan area.

Some police attended the meeting, unconvincingly dressed as peasants. They videotaped the proceedings, but so did GW representatives, anticipating that the PSB would edit the footage in order to make a case for closing down the discussion. GW submitted a summary report to the Yunnan provincial and the national Party discipline committees about what it had found out at that meeting. Once the meeting was concluded, the police informed the participants that they understood the issues but repeated that such proceedings were nonetheless illegal; then they transmitted their own report to the Yunnan provincial government.

A few days later, members of GW traveled to Xiaowan, just upriver from the Manwan Dam, to conduct a similar type of meeting. The local police, however, were prepared for them. GW officers appealed to the Xiaowan township government to ensure that the meeting would not be considered illegal, but the Xiaowan officials refused to meet with them. Again, showing Yu's penchant for "guerilla tactics," the meeting was held in the back yard of a farmer's house in an area large enough to accommodate the seventy or so people who had come for the meeting. There they discussed the issues and concerns of the local people regarding the Xiaowan Dam project until the meeting was broken up by the police. The Manwan Huaneng Power Company also sent representatives to the farmer's house in order to debate the issues. The farmers raised several concerns, but the Huaneng representatives claimed that they lacked the authority to discuss the matters raised by the farmers. Instead they gave vague promises about how the project would benefit the farmers in the long run while admitting that the short term would be difficult. After a while, the police arrived and "invited" Yu and others to the local PSB office. Eventually, the PSB let them go but did not allow them to stay in the area.

Peasant representatives from Dacaoshan (downriver from the Manwan Dam) attended the first Manwan anniversary meeting for 817 and invited GW to hold a similar forum in Dacaoshan. This group first went to the Dacaoshan county government and presented an introduction letter. But the government officials refused even to receive the letter and said that there was nothing for them to discuss. When the activists went to the Dacaoshan Dam site and to the surrounding villages, they were followed by two cars—one from the PSB and another from the local resettlement bureau—to shadow GW's activities.

By now, the Yunnan provincial government was also actively involved in monitoring these activities. It called an "urgent" meeting at which the governor and the provincial party secretary decided that GW should be closed down. They sought to find out if GW had broken any laws, and if they had, to use this as a pretext to dismantle the organization. The provincial government contacted GW and ordered its representatives back to Kunming immediately and informed the various other units in the entourage (i.e., journalists, film crews, and a social scientist from the Yunnan Academy of Social Sciences [YASS]) that they were forbidden to have any further contact with GW.

An investigation team representing the provincial government consisting of representatives from YASS, the Science and Technology Bureau, the Civil Affairs Office, and the PSB went to each of the previous locales to determine if GW representatives had broken any laws. They were confronted by the farmers who asked them why they were not investigating the problems of the local people with the same vigor. These organizations concluded that GW had done nothing illegal and, after conducting investigations throughout September 2004, issued a report to that effect. But they did note that GW had exceeded its organizational mandate of assessing rivers and were extending it into investigating the problems facing the *yimin*. The latter—a highly charged issue—is exclusively a government problem. Yet the conclusion that GW's actions were wrong but not illegal stood. One official on the investigation team representing the Civil Affairs Bureau went so far as to say that "we share [some of the concerns] with Green Watershed." Not long afterward, this official, who oversaw the management and registering of NGOs, was dismissed.

Then the investigation team visited the GW office, during which time Yu gave them a presentation with video clips during which he said that the Chinese Communist Party (CCP) wanted to help and emphasized the peaceful and constructive—and pointedly nonconfrontational—nature of the meetings. After this presentation, the members of the investigation team in attendance reportedly shook Yu's hand and told him that he was "doing the right thing." But there was one official who argued that GW should be inspected more thoroughly.

Nevertheless, the situation remained tense. Huaneng representatives informed SEPA that GW had been ordered by the Yunnan provincial government to disband. When GW representatives met with SEPA, the latter acknowledged the trouble that GW was in but said that there had not been enough evidence to that point to close down the NGO. Provincial officials informed Yu that he was "sitting on the edge of a cliff and that one mistake would cause him to fall over," after which they would shut down GW. At the same time, there seemed to be a lot of personal empathy and support for GW's activities among local officials, even in the face of intense political pressure: when a provincial official criticized GW in a magazine article, he also wrote that "we should learn from Green Watershed because they cross mountains, cross rivers, and go into remote areas to conduct investigations...what they do we cannot yet do."[3]

GW's strategy foreshadowed its actions to educate the villagers along the Nu River about the impact the NRP would have on their lives. Indeed, GW and other NGOs used the same strategies and tactics at both the national and local levels for the Nu River campaign. These included interaction with and education of local officials, aggressive exploitation of existing laws, and empowerment of the peasantry. This case demonstrates the degree to which local activism can mobilize those people directly affected by the hydropower dams to take action. However, although it illustrates intense local activity, it is insufficient to cause substantive policy change. In this sense, it has more in common with the unsuccessful protests at Pubugou described in chapter 3. It stands in sharp contrast to the far more successful campaigns of GW and others in casting the Nu River Project as the most controversial and high-profile debate over hydropower in China today.

3. This section draws from Interview 05KM03C, July 20, 2005.

The Battle over the Nu River, 2003–Present

The Nu ("Angry") River remains one of two undammed rivers in China.[4] Eventually becoming the Salween River in Myanmar, it is located in a remote stretch of western Yunnan province that hugs the contours of its border with Myanmar. The stunning beauty of the area is surpassed only by the almost unspeakable poverty of the people residing there. In fact, these are two of the more potent issue frames that shaped the locus of debate: economic development vs. environmental protection. This is due in part to the Yunnan government's scuttling of cultural heritage issues early in the debate, specifically in negotiating the precise contours of the World Heritage designation for the area.

On July 3, 2003, the Three Parallel Rivers (*san jiang*) region was approved to be a World Heritage site by UNESCO. However, there was something fishy about the Yunnan provincial government's strong interest in a clause stating that the World Heritage designation only comes into effect at an altitude of two thousand meters.[5] This curiosity was explained by the nearly simultaneous announcement that a gargantuan hydropower project involving thirteen major hydropower stations would be built along the Nu River and in other parts of the Three Parallel Rivers area. According to one (admittedly not completely disinterested) report:

> United Nations officials were puzzled when Chinese authorities asked that Tiger Leaping Gorge, one of the main features of the park, be excluded from the [designated protected area]. Why, the officials asked, was the magnificent gorge not to be included? "To allow for the construction of hydro dams," Liang Yongming [a professor at Kunming University of Science and Technology] told them.[6]

In fact, this public announcement was the culmination of intense behind-the-scenes activity by the proponents of the Nu River project. Although the hydropower companies are extraordinarily tight-lipped,

4. The Chinese character *nu* ("angry") is actually an approximation of the Nu (Lisu) minority's name for the river, which is "Nong." Richard Spencer, "China's Rivers to Be Dammed for Evermore," *Daily Telegraph*, January 20, 2006.

5. According to one source, this information has not been publicized because the provincial governor who negotiated that term is still in office. Although it remains an open secret, there is no paper trail on the subject. In any case, one need only look at a map of the UNESCO World Heritage Site for *san jiang* to see that it is not a coterminous area but is rather broken up into about eight pieces, each of which is presumably above the two-thousand-meter line. Interview 06BJ02, March 10, 2006.

6. Tashi Tsering, "Policy Implications of Current Dam Projects on Drichu-the Upper Yangtze River," available at www.tibet.net/tibbul/2005/0102/environment1.html, accessed May 27, 2007.

a journalist for the magazine *Economics* (*Jingji*) obtained a copy of the "China Hydropower Resource Picture Collection," compiled in 1991 by the Energy Department's hydropower development office, which called for six planned hydropower stations in the Nu River valley with a total capacity of 10,900,000 megawatts, or 10 percent of China's entire hydropower capacity.

In 1999, the National Development and Reform Commission (NDRC) decided to adopt the NRP based on its assessment of the energy situation in China and the requests of a group of representatives from the National People's Congress. Toward that end, the Water Resources Hydropower Planning Institute organized a bidding contest and decided on two planning units, the Beijing Survey and Design Institute and the Huadong Survey and Design Institute, to undertake the planning of the NRP. Their final design called for two reservoirs and thirteen dams.

Following the organizational restructuring of the water resources bureaucracy described in chapter 2, the Huadian Group got involved in developing the NRP at the beginning of 2003. It was decided that they would "develop both hydropower and thermal power generation, giving priority to hydropower." On March 14, 2003—several months before it was publicly announced or had even received the State Council imprimatur—Huadian and the Yunnan provincial government signed a letter of intent. According to the original plan, the power station at Liuku would be built first, with the construction to begin later that year.

In the spring of 2003, there was a meeting in Kunming to discuss the NRP that was attended by hydropower specialists from Beijing and Yunnan. The local experts all supported the project, while those from Qinghua University and the Chinese Academy of the Social Sciences expressed doubts or opposed the project outright. The views of the local experts—that is, the supporters—were carried in the local newspapers; the views of opponents were not. There was no firm conclusion at the close of the meeting.

At the same time, a television program, *Newsprobe* (*Xinwen Diaocha*), broadcast an expose that included interviews with Yu Xiaogang and with local officials in Nujiang prefecture. The interviewers asked these local officials—who had demonstrated their support for the NRP—some very basic questions about the NRP at Liuku. The officials betrayed an almost complete ignorance beyond the most general aspects of the project, underscoring the lack of information that was extended to local officials and their somewhat unqualified support for the project. Before the *Newsprobe* broadcast, prefecture-level people's congresses and political consultative

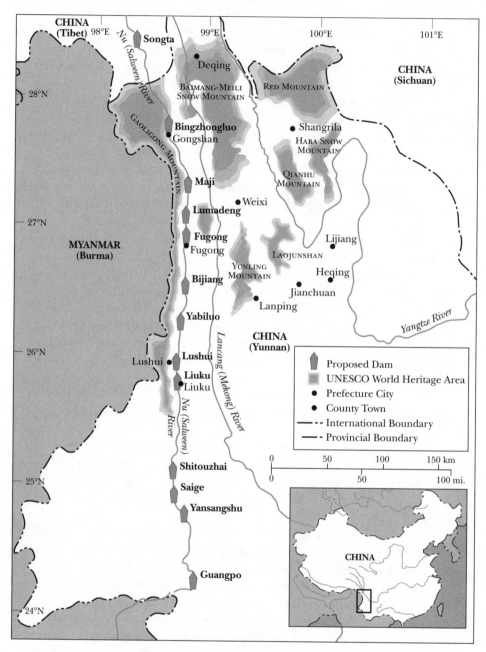

Figure 5.1 Map of the Nu River Project, http://internationalrivers.org/en/image/trd/109.

conference meetings had placed the NRP on the conference agenda. After the *Newsprobe* story, they were dropped from the schedule.[7]

The Opposition Mounts

The environmental activist Wang Yongchen learned about the NRP on August 16, 2003.[8] Prior to the Yangliuhu controversy documented in the previous chapter, Wang's only experience with dams was when she encountered the opposition of Thai villagers to a controversial dam project on a visit to Thailand in 2001. The person who tipped Wang off asked if she could provide a list of dam experts and other scholars familiar with the Nu River who might be able to provide information, i.e., ammunition, to the State Environmental Protection Administration (SEPA) for an upcoming meeting between SEPA and the NDRC. Wang provided the contact information for He Daming.

A famous river expert, particularly on the Nu, He is head of the Asian International Rivers Center at Yunnan University. He has been active in researching rivers in Yunnan for decades and has perhaps more information on the Nu River than any other single person in China. In the first week of September, He presented his opposition to the NRP at the "Nu River Valley Hydropower Development and Ecological Environmental Protection Issue Expert Forum" organized by SEPA in Beijing and attended by more than seventy experts (and ten journalists), although the content was characterized as an internal discussion (*neibu yantaohui*).[9] He was the first local scholar to oppose Nu River development. Nevertheless, although He's was a single, isolated voice, his opinion quickly caught on and snowballed dramatically.

On October 25, the NGO Green Earth Volunteers (GEV) organized a petition in which sixty-two people from the fields of science, arts, journalism, and grassroots environmental protection signed their opposition to the NRP at the second meeting of the China Environment and Culture Promotion Society. This petition was publicized through the media and elicited widespread public support in opposing the project. The battle lines were being drawn. An expert at the Chinese Academy of Sciences Finance

7. Interview 04KM07, August 24, 2004.
8. On the nature of the opposition, see also Ralph Litzinger, "In Search of the Grassroots," in *Grassroots Political Reform in Contemporary China*, ed. Elizabeth Perry and Merle Goldman (Boston, Harvard University Press, 2007). Litzinger's argument, shared by many of his colleagues in the same volume, that much of the opposition was kick-started by exogenous ("metropolitan-based environmental") forces is perfectly consistent with the argument being made here. Indeed, although he does not refer to them as such, they make up a subgroup of policy entrepreneurship. I only differ with Litzinger on the impact of transnational activism in that I do not find any close direct relationship between international activism and policy change.
9. Interview 04BJ02, August 2, 2004.

Institute asserted that although he supported environmental protection in the abstract, he was forced to conclude that hydropower might be the only way to lift the local people out of poverty and into the ranks of modern society. He added that environmental activists should be asking themselves: Who benefits more from environmental protection, the well-to-do in Beijing or the local people? In September and October, local experts in Yunnan and the Yunnan EPA all signaled their support for the NRP.

The first local activist to take on this coalition was, not surprisingly, Yu Xiaogang and Green Watershed. On October 1, 2003, he began his own survey of the Nu River valley. Yu's goal was to obtain an "objective understanding of the situation" and to influence the Yunnan provincial government as a disinterested NGO. But Yu also had an agenda. As discussed above, Yu had spent much time and energy monitoring the effects of the Manwan Dam along the Lancang River and anticipated many of the same problems encountered there—the adverse impact on resettled people, landslides, and other negative environmental effects—to be present in the area covered by the NRP. Yu had hoped to meet up with SEPA officials who had come to survey the Nu River at the same time, but SEPA abruptly revised their itinerary and returned to the provincial capital, Kunming.[10]

In November, the venue of the debate was once again in Beijing, specifically the "Third Meeting of China and the United States Environment Forum." There were a few NGOs in attendance. Wang Yongchen's group, GEV, and some others successfully pushed for a discussion of the Nu River despite unsuccessful appeals to UNESCO's Beijing office to take action. At the meeting, there was heated debate. On the one hand, some argued that it would be impossible for local residents to move out of poverty without the establishment of these dams. The other side countered that after the dam was built, local people might be forced to pay higher rates for electricity than their counterparts in urban areas. Moreover, there was concern about how these rural resettled people would be able to make a living once they were removed from their agricultural environment. Although these debates echoed earlier ones, the meeting was significant in that it led to a diffusion of opposition to the NRP throughout China's NGO network.

At around the same time, the World Rivers and People Opposing Dams meeting was held in Thailand. Among the participants were the Chinese activist NGOs GEV (Wang Yongchen), Friends of Nature (discussed in chapter 2), Green Island, and GW (Yu Xiaogang). At this meeting, NGOs from over sixty countries signed a petition to protect the Nu River and sent

10. Interview 05KM03C, July 20, 2005.

it to UNESCO. UNESCO replied by stating that it was paying close attention to the NRP. In addition, more than eighty NGOs in Thailand sent a letter to the Chinese ambassador in Thailand on the Nu River issue.

At the National Level

The National Environmental Impact Assessment (EIA) Law went into force on December 1, 2003. Importantly, it did not have a grandfather clause. In August 2003, barely a month after the Three Parallel Rivers region was declared a World Heritage site and right around the time that Wang Yongchen was getting wind of the NRP, the NDRC convened a meeting to examine the "Nu River Middle and Lower Reaches Hydraulic Planning Report" and approved it within two days, on August 14. Bolstered by this momentum, the Huadian Group rushed to get its NRP proposal approved by the State Council before the EIA Law came into effect, thus freeing them of the constraints of the law.[11] The proposal itself was "very simple" and contained no provisions on the impact of the NRP on the environment.[12]

The bare-bones proposal, combined with the speed with which Huadian sought to push it through the State Council approval process, raised the suspicions of some, especially at SEPA. Mou Guangfeng, vice director of the Environmental Impact Assessment Office and director of SEPA's Supervision Department, in particular, was troubled by this. Mou, described by *Southern Weekend* (*Nanfang Zhoumou*) as "a lone voice in the wilderness," sought the assistance of Wang Yongchen. According to the report, Wang recalled later that she encouraged the SEPA official, who was feeling isolated and powerless, by telling him, "SEPA must stand firm and never give up."

> The official suggested mobilizing experts on the Nu River to help mount a campaign. Through a massive effort by groups such as Green Earth Volunteers and the Yunnan-based Green Watershed, scholars, experts, citizens, and sections of the media rallied to the Nu River cause. For example, Shen Xiaohui, a senior researcher at the State Forestry Bureau, succeeded in submitting a petition letter to the National People's Congress and the Chinese People's Political Consultative Conference with the help of Liang Congjie, a CPPCC member and founder of the Beijing-based Friends of Nature group.[13]

11. Their fears were not unfounded. When the Three Gorges project was submitted to a pro forma vote at the National People's Congress in 1992, one-third of the delegates either voted against it or abstained, a stunning symbolic rebuke of the State Council and particularly Premier Li Peng. Antoaneta Bezlova, "Let Public See Secret Mega-Dam Plans, *Inter Press Service/Global Information Network*, Oct. 26, 2005."
12. Interview 05BJ02, July 4, 2005.
13. Deng Jie, "Environmental Protection's New Power is Growing," *Southern Weekend*, December 27, 2005.

Mou's opposition almost singlehandedly halted the momentum fashioned by Huadian and delayed the processes indefinitely, most importantly, beyond the critical date of December 1.[14] As a result, Huadian was forced to undertake an environmental impact assessment of the NRP and submit it for approval. Mou's resistance helped motivate others to signal their disapproval of the project. A delegate to the National People's Congress, He Shaoling, a senior engineer at the China Institute of Water Resources and Hydropower Research, went on record to the effect that "The Nu River Dam project must go through an independent and authoritative investigation before any decision on its future should be made. Not only is this in accordance with Premier Wen's call that China's development must be based on science, but it is the law."[15]

On February 18, 2004, Premier Wen Jiabao stated that "such a large hydropower station project that draws high social attention, and has environmental controversy, should be cautiously studied, and scientifically decided."[16] This effectively suspended the NRP. On April 9, 2004, Wen's decision was reported in the *New York Times*.[17]

Mou Guangfeng's opposition did not come without a price, however. His rivals waited for an opportunity to strike back. This apparently came after Mou made one comment too many to the press. He was fired from his position as vice director of the Environmental Impact Assessment Office at SEPA and almost immediately reinserted into his former position, but only in an "acting" capacity. In professional limbo as a *sunshi yuan*, Mou remained administratively active but was marginalized politically.

At the Local Level

These developments were not limited simply to the political process in Beijing. On February 13, 2004, at the Second Session of the Ninth Yunnan Provincial Political Consultative Conference, Dai Kang, vice head of the Yunnan Democratic Coalition Provincial Committee, questioned the NRP. This was the first time an opposing voice appeared within the Yunnan provincial government. A source present at the meeting told journalists that

14. It is rumored that a secretary to Premier Wen Jiabao has close connections to "somebody in the environmental community," which made it easier to get such issues on the table at the State Council. Interview 04BJ03, August 3, 2004.

15. Ray Cheung, "Assessment of Dam Project Urged," *South China Morning Post*, March 12, 2004.

16. Cao Haidong, "*Nujiang de minjian baowei zhan*" [The NGO Battle over Protection of the Nu River], *Economics (Jingji)*, May 2004.

17. Jim Yardley, "China's Premier Orders Halt to a Dam Project Threatening a Lost Eden," *New York Times*, April 9, 2004.

the Yunnan provincial leadership was embarrassed by this unexpected opposition. A week after the meeting, leaders of Yunnan province were sent to Beijing to study the central government's "sustainable development" proposals. Dai told journalists that GW's activities had inspired him to take such proposals seriously, adding that while many believe government experts to be heavily biased, nobody affiliated with the CCP was willing to speak up and express their true feelings on the issue and question the government and that the government should pay attention to their words. This, however, was the exception that proved the rule as the Yunnan provincial government successfully kept a lid on the opposition from within the government throughout the process.[18]

Indeed, the Yunnan provincial government was able to maintain a relatively solid united front throughout much of the process, including the support of the subprovincial governments. Although many have argued that these local governments were willing allies because of the rents they could collect, it would be inaccurate to attribute unqualified support for the NRP by all local officials to greed alone. In addition to the myriad considerations discussed in chapter 2, there is also general ignorance combined with a tendency to follow orders. For example, local generalists have been in favor of the NRP. They do not appear to have been involved in any of the actual planning, but according to one source, "they buy it." One compelling reason is poverty: Nu prefecture has been one of the poorest prefectures in Yunnan "since forever," and three of the poorest counties in Yunnan during the 1990s were within Nu prefecture. These officials are always trying to find a way out of poverty so that they can rely on local tax revenue and not always have to apply for money for projects—which often comes with strings attached. For example, logging was seen in the 1990s as a way out of poverty, but then the logging ban was enacted in 1998. For those interested in economic development, the potential posed by the NRP is huge, both for smaller-scale projects such as the mini–hydropower station at Dimaluo and larger projects such as the dam at Liuku. And, of course, economic development is an important dimension for career advancement decisions.

The village committees understand that village autonomy as written in the laws means that the land belongs to the village collective—not to the state—and that the village collective has the land use rights. But specific policies such as the logging ban, land slope conversion, and the NRP all demonstrate how the state can step in and override such autonomy.

18. Cao, "NGO Battle over Protection."

At the prefecture level, officials are for the NRP and are involved in some of the planning because they know the area and also because it is their administrative responsibility. But according to a source who lived in Gongshan county more than a year, these same officials are "absolute cretins" who "cannot possibly have been involved in the actual planning of such a technically challenging program." This suggests that there is a great deal of planning that is taking place outside of the realm of local politics and that it is undertaken without the active participation of these same local officials.[19]

In other areas, the picture is even more complex. Some officials within the Nu prefecture government are against the incursion of big hydropower projects. For example, in Wujiajia, the Nu Prefecture Hydropower Corporation and some of its Nujiang prefecture government patrons are against the NRP because part of it, the Maji Dam (located three kilometers from Wujiajia), will obviate—indeed, submerge—its own operations for a smaller hydropower station there. This latter project requires a major bridge; however, if it is submerged, which would happen if the Maji dam is to be built, then it will be a net loss to the local government. In addition, if the Maji Dam goes forward, it will affect the ongoing Dimaluo hydropower project. Eighteen families have already been resettled at the lower banks of the river for the Dimaluo project. If the Maji Dam goes forward, these people will have to be resettled yet again, something the local government does not relish.[20] In addition, at least one Gongshan official admitted in private that he couldn't care less about the NRP, and officials from the Gongshan Agriculture Bureau have mentioned that half the land in the area would be flooded by the dam just north of Maji; even with this high-quality land in circulation, the people are still fighting off hunger.[21]

There is also some variation across local administrative agencies, which stands along the various points of debate. The local EPBs in Gongshan county and Liuku municipality—that is, along the upper and middle reaches of the Nu, respectively—are in favor of the NRP, in part because they would prefer to see a larger project (presumably one that would be forced to adopt some "green" technologies) than the mushrooming of small hydropower projects that are largely unregulated and that are large contributors to local pollution.[22]

19. Interview 04KM07, August 24, 2004.
20. Interview 06BJ/KM01, March 18, 2006.
21. Interview 06FG01, March 13, 2006.
22. Ibid.

Digging In: 2004–2005

Rather than adopt a "wait and see" attitude, the players on all sides of this issue scrambled to increase their chances of victory; the two years from early 2004 to early 2006 were full of intense activity by the various groups involved. In March 2004, a journalist from *Economics* attempted to interview He Daming but was told that He probably would not be able to agree to an-views, in part because the Yunnan government had had a "talk" with him.

In February, the Twenty-Eighth session of the United Nations World Heritage Committee, which happened to be convened in China, issued Document WHC-04/28.COM/15B, which states that the World Heritage Committee

> [expressed] its gravest concerns on the impacts that the proposed construction of dams could have on the outstanding universal value of this World Heritage property; [invited] the State Party to respond to the calls of its academicians, conservationists and scientists consider letting the Nu Jiang River continue to flow naturally through and beside the World Heritage area; [and recognized] the importance of the energy sector in the development of the [*sic*] Yunnan Province and urges the Provincial and the Central Governments to seek alternatives to hydropower in order to ensure long-term protection of the Three Parallel Rivers of Yunnan Protected Areas which harbours the richest biodiversity assemblage in China and may be the most biologically diverse temperate ecosystem in the world.

The committee requested that China complete a report no later than February 1, 2005, "on or around the World Heritage property for examination by the Committee at its Twenty-Ninth session in 2005."[23]

Throughout this period, other NGOs actively publicized the Nu River Dam story. In February 2004, a group of twenty journalists, environmental protection volunteers, scholars, and other experts, led by Wang Yongchen, traveled to the Nu River for a nine-day survey motivated in part by the criticisms of dam proponents that environmental people had not actually visited the Nu River Valley itself but were simply lecturing from an implied "ivory tower" perspective in Beijing. This group traced the route of all thirteen proposed dam sites, traveling incognito as "tourists" in order not to arouse suspicion. On their return, the group presented a photo exhibition of the trip in Beijing, with Wang reportedly dipping into her own savings to help underwrite

---

23. United Nations World Heritage Committee, "Decisions Adopted at the 28th Session of the World Heritage Committee" WHC-04/28 COM/26, Suzhou, China, 2004, 77–78.

the costs. On March 14, one week before the opening of the exhibition, the group launched a Chinese- and English-language website on the Nu River.[24]

Meanwhile, Shen Xiaohui of the State Forestry Bureau began preparing two proposals for an upcoming Chinese People's Consultative Conference meeting, based on his observations on the nine-day survey trip, entitled "Protect the Natural Flowing of the Nu River, Stop Hydropower Development" (*"Baohu tianran dahe Nujiang, tingzhi shuidian tiji kaifa"*) and "Proposal for Classifying Rivers and River Valleys, Coordinate Ecological Protection and Economic Development" (*"Guanyu fenlei guihua jianghe liuyu, xietiao shengtai baohu yu jingji kaifa de ti'an"*). At the meeting, Liang Congjie, founder of FON, put forward Shen's proposals in lieu of his own, demonstrating the increasingly porous interface between NGOs and representatives of some government organizations.

Around the same time, in Shanghai, an anonymous official in Huadian said that "neither the instruction written by Wen [Jiabao], nor any file containing the destiny of the project has reached us so far."[25] Indeed, dam developers continued to take hydrological and geological measurements at Yabiluo and build roads to a proposed dam site at Liuku. This was not a violation of Wen's directive per se because it only mandated the cessation of "actual project work" and did not cover "preparatory work," which continued apace.[26]

On April 16, the group that had undertaken the survey trip to the Nu River in February returned to the area, traveling to Wawa village, Fugong county, an area with a large concentration of minority people. The party secretary of Pihe prefecture said that one of the thirteen hydropower stations was planned for the region. A villager named Pudayi pointed at the holes drilled into the rock walls along the banks of the Nu and said that they had been anxious about their future ever since the drilling started. He and the other villagers were particularly worried about the resettlement plans; according to the plan designed by the Beijing and Huadong Survey and Design Institutes, 48,979 people—about 5 percent of the number displaced by the Three Gorges Dam—would be relocated after the thirteen dams were to be built. Pudayi articulated the same sentiments as the peasants I interviewed in the area:

> We are in the smallest of worlds. No one interferes with anybody else. And we are not intruded upon. We live a peaceful life. Everyone has enough to

24. "Our Attachment to Nujiang River," available at www.nujiang.ngo.cn, accessed May 29, 2007.

25. "'No Conclusion Drawn' regarding Dams on Nu River," *Interfax*, April 14, 2004.

26. Ray Cheung, "Proposed Dam May Leave Poor High and Dry," *South China Morning Post*, June 14, 2004.

eat and enough clothes to stay warm. We can tell when it rains and when the skies are clear. We have our own lifestyle, customs, and festivals. What can we do if we are moved?

The village chief, Sangpuyi, added that he did not wish to move but that the villagers' words did not count. Pudayi summed up the situation: "Rich or poor, we want to live here. But if our country really wants us to move, we can't violate policy." These events seemed to have the desired effect, as expressed in the reaction of the vice general manager of the Huadian Group, Zhang Jianxin, who complained that public opinion had imposed delays on the entire NRP, which was affecting the Liuku power station in particular.[27]

These developments coincided with a press crackdown. In April, there were reports of a "news blackout" on the NRP, in which even the more aggressive media outlets such as *Southern Metropolitan Daily* (*Nanfang Dushi Bao*), *Southern Weekend* (*Nanfang Zhoumou*), and *21st Century Economic Report* (*Ershiyi Shiji Jingji Baodao*) stopped reporting on the issue.[28] This may have been partly due to the prison sentences issued to *Southern Metropolitan Daily*'s former general manager, Yu Huafeng, and its former editor in chief, of twelve and eleven years, respectively, on corruption and embezzlement charges, although many believe that they were a reprisal for the Southern Media Group's muckraking journalism. Interestingly, however, many other journalists and intellectuals issued a bold petition that criticized the sentences and decried the "illegal use of all kinds of measures, including juridical methods, to limit press freedoms and crack down on the media and limit its space."[29] Meanwhile, a spectacular piece of guerilla theater was being organized by Yu Xiaogang.

Making Movies

For years, Yu Xiaogang had been following the developments along the Lancang River closely. One of his principal concerns was local corruption relating to compensation for the 7,500 people resettled by the project. The Manwan hydropower station was hailed as one of China's "five golden flowers" during the Seventh and Eighth Five Year Plans because of its high

27. Pubugou Interviewee, July 8, 2005.

28. *Southern Metropolitan Daily* was among the first newspapers in China to break the SARS story in March 2003; it also reported in December 2003 about a possible recurrence of SARS. *Southern Weekend* reported—among many other things—that eight graves found in the Three Gorges Reservoir area were contaminated by anthrax.

29. Howard W. French, "China Tries Again to Curb Independent Press in South," *New York Times*, April 15, 2004; and Kelly Haggert, "No Nu News as Beijing Cracks Down on Crusading Papers," *Three Gorges Probe*, April 16, 2004.

没有补偿 那你们现在靠什么生活
人家（电厂）补偿叫政府吃了
就靠捡一点垃圾 娃娃卖工

Figure 5.2 Still image, *Nujiang zhi sheng* ("The Voice of the Nu River"), a DVD documenting Yu Xiaogang's May 2004 trip to the Manwan area.

profit-to-cost ratio. Yu found that one of the reasons for this cost efficiency was that many of the people who had their fields submerged had been inadequately compensated (each person received 3,000 RMB, or US$360) and were eventually forced to make a living collecting garbage.

As the Nu River Project stalled in Beijing, from May 25 to 28, 2004, Yu chartered a bus and brought a group of community leaders from Xiaoshaba village, Liuku, to visit the Manwan and Xiaowan power stations along the Lancang River in order to show these people what they might expect in terms of compensation from the Nu River Valley developers and the local government.

Yu and his film crew went to the village and stayed there for three days. They made the pitch to take the villagers to the Manwan Dam site only on the second day—that is, after they felt that the villagers were more comfortable with the outsiders. After discussing the matter, the villagers came back the following day to say that they agreed to go. Sensing the political sensitivity of the trip, the village chief let it be known that he would be in town buying a car and did not want to be bothered; he sent his brother in his stead. Similarly, the village Party secretary asked his father, a former village chief, to go in his place. The people that went were mostly volunteers who tended to be more active in village affairs.[30]

30. Interview 05KM03B, July 14, 2005.

At one point, after they had reached the Manwan area, their bus stopped when they saw people along the road sorting through garbage. They filmed an interview with an old woman from the area.

Q: Which village are you from?

A: We are from Tianba village.

Q: What are you picking through?

A: We pick up anything. Coke cans....

Q: What is the use of this stuff?

A: We sell them. As for this fish, I'll bring it home.

Q: Do you still have land?

A: No. no farm land, no land.

Q: Where did it go?

A: The land has been taken by the country. Farmland and land.

Q: Did you receive compensation?

A: No.

Q: No compensation, then what do you live on?

A: The compensation for the hydropower station was eaten by the government [*zhengfu chi le*]. Now we live by picking trash, and our kids are forced to work.

Q: How much can you earn in a day by picking through trash?

A: On good days, about 1 RMB (US$0.12); on bad days, practically nothing.

Q: The trash is so dirty, aren't you afraid of getting sick?

A: No, we are brave.

Q: Where is your home?

A: [Pointing] Over there. The place to which we originally moved to had a landslide, so we had to move again. Then another landslide, so we have had to move yet again.

Q: Do you feel your life was better before you moved or better after you moved?

A: If I told you, I'd be in violation of the law. Before the power station was built, our life was pretty good. We had farmland, four and a half *mu*. It could yield over 6000 kilograms a year. Now, we have nothing. No farmland, no land, nothing.

Yu had expressed surprise that the *yimin* were still sifting through garbage near Manwan, as he had been there two years earlier and would have expected the local government—if only for reasons of public relations—to have created a sort of Potemkin village to show outsiders that the locals' socioeconomic standing had improved. The people who went on the trip were impressed by what they saw, but they also came under increased political supervision, with local officials telling them once they had returned that Green Watershed was an "illegal agency" and thus their trip had been an "illegal act."

This entire trip was filmed and committed to video compact discs (VCDs) and DVDs. These discs have been distributed by peasant activists throughout the province and among NGO circles in Beijing (the filmmakers acknowledge that this film could never be broadcast on state television because it is too provocative). The spectacular landscapes of the Three Parallel Rivers region are matched only by the riveting face-to-face interviews with local peasants. In particular, the images of broken peasants crying during interviews and picking through garbage are very powerful, making the documentary extraordinarily effective propaganda.[31]

A Controversial Precedent at the UNHSD Conference

An important milestone, in terms of process if not outcome, occurred from October 27 to 29 at the United Nations International Conference on Hydropower and Sustainable Development (UNHSD).[32] The meeting left a mixed message. On the one hand, although the vice minister of the NDRC, Zhang Guobao, acknowledged the "indispensable contributions" of the opposition, he nonetheless maintained that China would not abandon hydropower development and that China would move forward on that front.[33] One source told me a couple of months earlier that there had been attempts by the Water Resources Hydropower Research Institute (*shuili shuidian yanjiu yuan*) to "hijack" the conference.[34] In short, the verdict of the meeting was that the thirteen dams of the NRP would not adversely affect the environment.[35]

However, perhaps buoyed by the momentum of his "field trip" to the Manwan site with the villagers from the Nu River Valley, Yu Xiaogang provided another shock. Breaking all of the unspoken rules of such a conference, Yu brought with him two actual peasant activists—people traditionally excluded from such meetings for policy experts and officials—and saw to it that one of them, Ge Quanxiao, addressed the audience directly.

Ge's speech, "The Relationship between Dam Construction and the Rights of Original Inhabitants to Participation," addressed the case of the dam at Tiger Leaping Gorge along the Jinsha River. However, substantively it echoed many of the concerns surrounding the Nu. Starting in a folksy man-

---

31. So effective, in fact, that this video was singled out during a crackdown on Chinese NGOs in December 2005; it is now forbidden to distribute it.

32. Interview 05SG01, July 18, 2005, provided additional information for this section.

33. "China Set to Press on with Hydropower Development despite Environmental Concerns," *Interfax*, October 28, 2004.

34. Interview 04KM08, August 24, 2004.

35. Interview 05KM02A, April 27, 2005.

ner, Ge introduced himself as a "common resident" alarmed at the prospects of having his community's farmland submerged and up to a hundred thousand people in the region displaced. "As the biggest group of stakeholders, I think that we local people must participate in the decisionmaking process for the sustainable development of this beautiful place," he said.[36]

Ge continued by citing the various laws and regulations that officially protect the rights of villagers. He started with the State Council's "Suggestions on the Establishment and Improvement of the Democratic Administration System for Village Affairs," published in the *People's Daily* on June 12, 2004. This document focused on the "four rights" of farmers—the right to be informed, the right to participate, the right to participate in decisionmaking, and the right to undertake monitoring. Ge argued that hydropower policy, as it was currently being pursued, violated these rights. To counteract the withholding of information and to pursue the peasants' "rights" to access of such information, Ge and others "collected and distributed materials about national policies, academic debates and media reports to farmers. Our purpose was to make the public know the reality in order to facilitate public participation in the discussion of the dam construction." Ge argued that villagers' goals diverged from what the dam proponents had said they were: "To common farmers like us, tall buildings and the prosperity of cities are not our goals. What we need is bright sunshine, fresh air, clean water, and fertile lands. We have wealth both emotionally and in substance within this beautiful natural environment. We are enjoying the comfortable life that are the fruits of the policy of development."

Ge spoke indirectly to the importance of the media and transparency, without which domestic and international experts cannot debate these issues. He continued to argue that the preparations for the dams were having a negative impact on the ongoing, more modest infrastructure projects that were actually benefiting local residents: "Infrastructure constructions in this region were all cut, supportive projects for farmers were transferred to other sites, country roads were changed to temporary pathways for resettlement, and the position of Jinsha township leader has been empty for more than half a year because no one is willing to take over the position." Ge bemoaned the loss of traditional courtyard houses and the social upheaval that would undermine the "harmony" of ethnic groups that evolved naturally over generations in the region. He went on:

> Therefore, I believe that the protection of World Heritage of "Three Parallel Rivers" is not only a task of river protection but also involves the protection

36. See www.irn.org/basics/conferences/beijinghydro/pdf/dampar.pdf, accessed July 27, 2007, 1.

of the stable living environment of local inhabitants. The value of world heritage protection should be based on the harmonious development between nature and human beings and harmony among all ethnic groups. This should be the real meaning of protection. Through discussions and communication, farmers living in this region believe that in the residential places of multiethnic groups, the construction of the dam will disintegrate the harmony of minority ethnic groups, which will undermine national policies for ethnic groups.[37]

Finally, Ge moved to the issue of resettlement. Although he acknowledged the complexity of this issue, the most important part of it is the measure of the productivity of the (resettled) land: the former land farmed by peasants has been farmed for a reason—it is fertile. The new land that has been allocated to the *yimin* is also available to them for a reason—nobody else wants it. Ge said that the *yimin* are "disadvantaged groups, not 'cultivation troops' supported by government finance."[38] Another important part of the issue, argued Ge, is that peasants who are resettled and become urbanized often do not seem to fit in and naturally gravitate to the margins of urban society, reduced to begging or engaging in criminal activities,[39] a point acknowledged by an official charged with resettling the *yimin* displaced by the Three Gorges, whom I interviewed in late 2005.[40]

Ge concluded by laying out several suggestions, including the establishment of a strategic planning system for dam projects that would give more weight to environmental and social impact assessments, in which the rights and interests of farmers in the dam areas can be expressed completely. Veto rights should also be included within the assessment procedure to "avoid disorder during the planning stage." Second, Ge suggested that "the administrative departments, local governments, and investors should adopt the report of the World Commission of Dams 'Dams and Development—A New Framework for Decision-making' for the policies surrounding dam construction. The purpose of the suggestion is to facilitate the collaboration between Chinese experiences and international advanced methods and opinions."[41]

The authorities were reportedly furious with Yu Xiaogang for "unleashing" Ge at the conference. Ge's clear articulation of the issues from the vantage point of someone with a "worm's-eye view" provided a degree of

37. Ibid., 3.
38. Ibid., 4.
39. Ibid., 2.
40. Interview 05ZQ01, November 5, 2005.
41. See www.irn.org/basics/conferences/beijinghydro/pdf/dampar.pdf.

credibility that was very difficult to dismiss, especially as it was published in the documents of the official proceedings of the conference. This, of course, guaranteed that Ge's remarks would reach a wide audience, far beyond the conference participants.[42]

In a twist that characterized the entire process, although the verdict of the conference was that there would not be significant environmental degradation, the immediate construction on the NRP was opposed, but this time by the NDRC! This effectively halted the project. Indeed, Huadian was so sure that the verdict would be positive that they had started work again on the project, only to bring it to a halt again.

In December, in an attempt to overcome the stalled project, the Yunnan provincial government held a meeting with various experts in order to convince them that the project was feasible and environmentally friendly. Nujiang prefecture was in agreement, the Yunnan government was in agreement, and so, apparently, were some of the experts. The provincial EPA was in favor, even though SEPA was not.

Then, early in 2005, the UN sponsored a meeting in Beijing on dams and development. Their conclusion was that Liuku could establish some sort of water recycling operation (generically referred to as "*zaisheng*") in which water would then somehow be poured back into the river. In this case, Liuku could go ahead, went the thinking, but the other twelve hydropower stations could not.

But the plot thickens. A vice director of the NDRC, Zhang Guobao, pushed experts to come up with plans to make less of an environmental impact. Perhaps this is why the NDRC called a halt to the project in December: to regroup and to establish an environmental impact assessment plan that would be impossible to oppose.[43]

By January 2005, there were conflicting reports over the status of the NRP. The deputy director of SEPA, Pan Yue, announced that public hearings—extremely rare in China—would be held on the project. While SEPA officials admitted that they did not have the power to stop the dam project singlehandedly, they pinned their hopes on the EIA Law as a means by which they could get additional leverage to compensate for their relative power inadequacies.[44] At the same time, at the National People's Congress

---

42. Liuku villager He Lixiu was invited to a conference in Beijing, but the day before she was scheduled to leave, ten plainclothes police officers descended on her home and told her that if she attempted to travel to the conference, she would be arrested. Richard Spencer, "China's Rivers to be Dammed for Evermore," *Daily Telegraph,* January 20, 2006.

43. Interview 05KM02A, April 27, 2004.

44. " 'No Conclusion Drawn' regarding Dams on Nu River."

in March 2005, the minister of water resources, Wang Shucheng, announced that four of the dams would move forward. Indeed, basic surveying around Liuku—the first of the thirteen dam sites to be built—was continuing, albeit in a roundabout fashion, as the order to cease and desist had been interpreted to mean that construction of access roads, geological surveys, and the insertion of marking pegs into the hillside could go ahead.[45]

In June there was a meeting of (pro-dam) experts in Yunnan, and they issued a report to Wen Jiabao about the situation regarding the Nu. The response from the State Council was that the report needed to be revised (*tiaozheng*) without stating exactly how. This has led to differing interpretations of the State Council decision: the Yunnan provincial government has argued that it is OK to go ahead with the project (and continue with the related infrastructure development) pending some revisions, while SEPA took this decision to mean that the project should remain on ice until it is revised (and presumably canceled if it is ultimately unable to be revised adequately). Also in June there was another meeting in Yunnan, where it was decided that even if the State Council ultimately says no, Yunnan would move ahead anyway and get money from the Bank of China and other sources.[46]

In July, UNESCO recalled its decision 28.COM/15B.8 from the previous spring and commended "the State Party for responding to its request and for submitting the Management Plan for the World Heritage property, as well as for its efforts to conserve the property and the recent progress." Although it also noted that progress was slow in establishing a management regime for the area due to a lack of funding, it was clear that UNESCO was not going to deliberately risk a clash with Beijing.[47]

Fighting Words: Fang Zhouzi

The other side similarly escalated the conflict, ratcheting up the rhetoric to undermine the credibility of the hydropower critics. Insofar as policy entrepreneurs' credibility is their most important asset, this counteroffensive was an important strategic decision by the pro-hydropower forces. They did not hold back, fueling an intense academic debate by intellectuals-for-hire on both sides of the issue. The lightning rod for the debate is a scholar named Fang Shimin, better known by his nom de plume, Fang Zhouzi, a

45. Hamish McDonald, "Crouching Tiger, Hidden Power," *Sydney Morning Herald,* March 26, 2005.

46. Interview 05KM02B, July 20, 2005.

47. UNESCO, "State of Conservation of Properties Inscribed on the World Heritage List," WHC-05/29.COM/7B.Rev, 14.

biochemist and physicist with a doctorate from Michigan State University, who has become a vitriolic critic of those who, in his view, do not reason rationally and scientifically.[48] Something of an apostate, he was originally a critic of the regime but has more recently become something of a pit bull in defense of its policies. His polemics are extremely entertaining—direct, confrontational, peppered with ridicule—and also quite effective.[49] Not surprisingly, the main targets of his criticism have been those most active in their opposition to these dam projects.

Fang's principle theme is the distinction between "rationalism" and "emotionalism" or "enthusiasm." After the Indian Ocean tsunami of 2004, He Zuoxiu, a member of the Chinese Academy of Sciences, wrote an article entitled "Humans Need Not Worship Nature."[50] Wang Yongchen's rebuttal in the *Beijing News* (*Xinjing Bao*) was called "Worshipping Nature Is Not Anti-Science." This opened up the lines of debate, and in the first six weeks of 2004, no less than thirty-one articles were published that took one side or the other. Fang took exception to what he saw as an obvious bias: "Of the twenty-two authors of the thirty-one articles, only five authors and nine articles are in opposition to nature worship, with one being neutral, while the rest are all for worshipping nature." Indeed, he argued that the major opponent of "nature worship," Zhao Nanyuan, was unable to get his articles published in the *Beijing News*. He went on to say that those who worship nature tend to have a liberal arts background and those who do not come from scientific educational backgrounds.

The former group tends to anthropomorphize nature, as when Wang Yongchen described the mountain god's anger when describing the deaths incurred on a Chinese-Japanese joint climbing expedition on Meili Snow Mountain (*Meili xueshan*). Far more important, however, this discussion over "nature worship" is a proxy for the debate between "real science" and "fake science" and the far larger question about what the proper contours of the environmental debate should be. This debate is not over the importance of environmental protection but rather over what the nature and purpose of environmental protection should be. Those who fall into the

48. His website is *New Threads*, www.xys.org. The title of his most well-known book is *Kuiyang: zhimian Zhongguo xueshu fubai* [*Ulcer: Confronting China's Academic Corruption*], Hainan chubanshe, 2001.

49. On the NRP, he has written, "Of the planned thirteen dams on Nujiang, some are controversial, some are not. Can't we start building the noncontroversial ones?" Fang, *Nujiang xiu ba yu xingtai ruhe quanheng?* ["Travel to Yunnan: What is the Trade-Off between Building a Dam and Ecology?"] www.bioon.com/popular/argue1/230365.shtml, accessed July 30, 2007.

50. "Worshipping nature" (*jingwei daziran*) became a popular topic of discussion after the SARS crisis of 2003.

"nature worship" camp say that environmental protection should benefit the environment and those who are opposed feel that people should benefit most from environmental protection—the main goal of environmental protection is to combat pollution. The latter group argues that science—not emotion—should guide environmental protection, while the "nature worship" camp is not immune from falling into superstition or ignorance, as they lack scientific training.[51]

According to Fang, Wang Yongchen is emblematic of the "nature worship" side. In responding to a debate over semantics, Fang argued that

> The exact meaning of "*jing*" is not important. In Wang's mind, it means "worship." Her words at tom.com that the "Nujiang is calling for help from us humans" advocated that Chinese people should worship nature and fear it as a belief.... The gist of Wang's "nature worship" lies in fearing nature, and preventing humans from learning from it, thus being able to utilize, and change nature. This is in conflict with scientific thinking, research and utility.[52]

Elsewhere, he elaborated on this point:

> At its heart, "nature worship" is fear. This is the same as then ancient people's pantheism, or the interaction between the heavens and Man. They all treat nature as god that has consciousness and will exact revenge if it is offended.... This is an irrational and uncivilized belief. It is the opposite of scientific thinking, because modern science has a basic assumption that the materialistic world is an objective world, without consciousness. Natural principles are not affected by Man's will. To ask people to stay at a respectful distance from nature, to oppose understanding nature through the scientific method, and to oppose exploiting and altering nature via the scientific method is anti-science.[53]

But perhaps the most incendiary moment was Fang Zhouzi's speech at Yunnan University on April 8, 2005. After some preliminary remarks in which he criticized Wang Yongchen for a significant misreading of the proposed power-generating capacity of the NRP, he moved to his principal theme and target: a full-on assault on He Daming. Recall that He Daming

51. Fang, *Renlei shi fo ying jingwei daziran lixing yu riqing shu chong?* ["Should Humans Worship Nature, What Is More Important: Rationalism or Enthusiasm?"], env.people.com. cn/BIG5/35525/3210966.html; and *Women yao you shenmeyang de "huanbao"?* ["What Kind of Environmental Protection Do We Want?"], tech.sina.com.cn/d/2005-02-23/0837532641. shtml, both accessed July 30, 2007.

52. Fang, *Jingwei daziran shuo shi zai xuanyang tian ren ganying de guannian* ["Worshipping Nature Is Promoting the Interaction between Heaven and Humans"], tech.163.com/05/0121/09/1AK2CGLB00091537.html, accessed July 30, 2007.

53. Fang, *"Jingwei daziran" jiu shi fan kexue* ["Nature Worship Is Anti-Science"], tech.163. com/05/0121/09/1AK2CGLB00091537.html, accessed July 30, 2007.

had opposed the NRP from the beginning and is perhaps the most knowledgeable person in the world on the Nu River Valley. People at Yunnan University told me that he has been under tremendous pressure to keep quiet, and he has been extremely careful not to take any unnecessary risks, such as talking to the press. Moreover, they said that the Yunnan authorities had planned for this event as a way to chip away at He's credibility. Those present with whom I spoke were simultaneously amazed and outraged by the brazen nature of the attack.

Fang ridiculed He Daming and his coining of the phrase, "original ecological status river" (*yuanshengtai heliu*) and He's argument that the Nu River falls into this category. "Is [He] making up a new concept so that he can be dressed up as a founder of a new science, and then apply for research funds from the government?" asked Fang. In fact, he went on, there have been all sorts of man-made projects on the Nu and its tributaries (although the ones he cited were in Myanmar). When he visited the Nu, said Fang,

> I did not see much primitive forest....I have seen patches of farmland on quite steep slopes. Locals call them "big character posters on the wall" [*gua zai qiang shang de dazibao*]. Slash-and-burn is the method used to farm such land....We think that's a very primitive and outdated method, but the head of Nujiang prefecture told us this is the most progressive method that suits the local condition. Why? Because the slopes are too steep to plow. And the soil along Nujiang is poor. Local people cannot afford to buy fertilizer, so they can only use fire to burn wild grass. And such land can't be used after a few years, so they have to burn another patch. Moreover, the production of that kind of land is very low: 40 to 50 kilograms per *mu* is quite good. To support one person, five *mu* must be farmed.

In addition to evidence of slash-and-burn agriculture, Fang

> also saw many landslides and mud- and rock flows. Lu Youmei [CAS] mentioned that the cause was soil erosion brought about by farming. In slash-and-burn agriculture, another reason for this loss of soil and water is the excessive cutting of trees as a result of local people's need for firewood. These local people use firewood for heating and cooking. They told me that on average a family will cut 3.5 cubic meter wood annually....Yet another reason is the Minxin project—connecting villages with highways. Each highway destroys some farmland. So, during my trip, I did not see any sort of "original ecological status," regardless of how you want to define it. So, this "original ecological status" concept is purely made up for media, for getting research funding.

He went on to deride another of He Daming's terms: "vertical forest and valley area biodiversity" (*zongxiang lingu diqu*). After asserting that he had never heard the term before—"another new term for funding purposes"—he asked the audience sarcastically, "Is Professor He going to create 'forest and valley-ology' [*lingu xue*] and become the founder of that research program?"

Fang finished by saying that he himself is not a blind supporter of dam construction but that his opponents were against any and all dam construction. The NRP was not in any way a unique case; it was a dam project like many others. The problems associated with dams cannot all be resolved immediately, he argued, but that does not mean that all dam-building must cease until solutions to all these problems are found. Moreover, he maintained that his critics "used bullshit" to "speak outside their area of expertise." This was a serious problem because, he argued the media was being controlled by the anti-dam forces. He finished by plugging his own NGO: "my '*xin yusi*' [New Threads website] is a genuine NGO. It can act as a supervisor of the media in this debate."[54]

Far from dismissing Fang's criticism, a government-relations officer at a Chinese NGO told me that although he was furious with Fang for the content of his Yunnan University speech, Fang's criticism of emotionalism among many environmental activists is not without merit. In fact, he found himself having to distance himself from previous allies in order to maintain his own (and his organization's) credibility vis-à-vis the government officials with whom he interacts. Nevertheless, a large part of the problem is that the most basic information is often classified or otherwise impossible to obtain. As a result, dam opponents have to make do with anecdotal or impressionistic information, which plays right into the hands of their opponents, including Fang:

> The NGOs try very hard to get information, but it is difficult to do so. This can make for a one-sided debate. For example, "biodiversity" makes for a nice sound bite in the debate as some say that dams are bad for biodiversity and others say that dams are good for biodiversity. But how do you actually *measure* biodiversity? If you cannot, you are being "emotional" if you oppose dams. If you can, it is easy for opponents to say that your numbers are way off. But this is because some of the most basic information is confidential. Often, experts are afraid to share their information, so they refuse because if they do share it, their government-sponsored funding will be cut off.[55]

The overall result is that something as seemingly objective as scientific debate quickly becomes politicized, rendering it largely meaningless. The political dimension of how an issue is framed, therefore, takes on far more

---

54. Fang, *Fang Zhouzi zai Yunnan daxue de yanjiang: zhi ji wei huanbao fan ba renshi* ["Fang Zhouzi's Frank Critique at Yunnan University against Those Who Espouse Fake Environmentalism and Oppose Dams"], April 8, 2005, available at tech.sina.com.cn/d/2005-04-11/1357577996.shtml, accessed July 30, 2007.

55. Interview 05KM06, July 26, 2005.

importance than a simple debate over the factual merit of the project being discussed.

The Momentum Continues to Build: Late 2005

On August 25, an open letter signed by sixty-one organizations and nine individuals called for the environmental impact assessment of the NRP to be made public. However, this particular appeal raised the stakes enormously precisely because it did not stop with the NRP: "The signatories of the open letter would like to see the Nu River scheme become a test case, and the first of many such projects to be subjected to public scrutiny." Indeed, in the actual body of the letter, they wrote that the NRP "is not an individual case....We hope the process can help develop a set of science-based and democratic decision-making mechanisms, in order to cope with the over-heated and unregulated hydropower development boom."[56] By September 27, the number of signatories increased, as twenty-three new groups and 232 individuals further "endorsed the appeal, bringing the total number of signatories to eighty-four organizations and 331 individuals."[57]

But by the fall of 2005, some reports suggested a power struggle taking place by proxy through a highly politicized intellectual discourse within official channels of communication. In October, at a forum on hydropower in Beijing, Chinese Academy of Sciences (CAS) fellows He Zuoxiu and Lu Youmei (former chief of the Three Gorges project) argued that plans for the NRP should move forward. But an editorial in *China Youth Daily* attacked He's "illogical" views: "As for the impact of hydropower development on the environment...we would rather trust the conclusions of environmental departments." Citing Qinghua University professor Li Dun, the editorial argued that the only actors that benefit from these hydropower projects are the hydropower companies themselves, citing problems that arose from the Manwan Dam project.[58]

Closing out 2005, an expert panel was convened by the NDRC in Beijing on November 13, but key members of the opposition did not receive their invitations until two days before the panel was to begin. Moreover, the discussion was brief: "At the session, the experts were shown an environmental

---

56. Kelly Haggert, "Open Letter Puts Pressure on Beijing over Secretive Dam Plans," *Three Gorges Probe,* September 7, 2005.

57. Haggert, "Nu River Campaign Gathers Steam," *Three Gorges Probe,* September 27, 2005.

58. David Stanway, "Press Criticism of Nu River Hints at Deeper Power Struggle," *Interfax,* October 25, 2005.

impact assessment produced by the East China Survey and Design Institute and the Guodian Corporation's Beijing Survey and Design Institute, but the material was whisked away as soon as the panel ended. [One professor] joked that it had all happened so quickly and bizarrely that he couldn't even recall the name of the panel."[59] Huadian was still seeking a closed-door meeting with the NDRC and the State Council without having to go through the National People's Congress and without public disclosure of the environmental impact assessment report.[60]

The Debate Enters Its Third Year: 2006

Amid whispers and cautious optimism by the dam opponents, January 2006 saw the leaking of some of the main conclusions of the NRP Environmental Impact Assessment report. As with so many of the twists and turns of the Nu River saga, this latest development is open to interpretation:

> Chinese government environmental review has recommended reducing the number of dams included in a hydropower proposal on the Nu River in southwestern China to limit environmental damage and decrease the number of people who would be resettled, according to a Hong Kong newspaper report and a provincial environmental official.... The newspaper, *Wen Wei Po*, which has ties to the Chinese Communist Party, reported Wednesday that the recommendation calls for four hydro dams instead of the 13 in the original Nu proposal. The article, citing an unnamed source "close" to the government review, said a reduced number of dams would meet "the needs for economic development and environmental protection."... Officials with the State Environmental Protection Administration, which has responsibility for the review, declined to comment. But an environmental official in Yunnan Province, where many of the proposed dams would be built, confirmed that the environmental assessment review has been completed and recommended only four dams.[61]

This would not necessarily be the final word, however:

> If so, the project would next be presented to the National Development and Reform Commission, a powerful government ministry, for approval. Then, if approved, it would be presented to the State Council, or the Chinese cabinet, for final consideration.... The possibility of a scaled-down project would seem to represent a partial victory for environmentalists, academics and others who have lobbied against the project. But many environmentalists are concerned that approving four dams would merely serve as an

---

59. Deng, "Environmental Protection's New Power is Growing."
60. Bezlova, "Let Public See Secret Mega-Dam Plans."
61. Jim Yardley, "Hydroelectric Project in China May Be Shrunk," *International Herald Tribune*, January 12, 2006.

opening for the full project to be built later.... The environmental assessment report cited by the Hong Kong newspaper has itself become a point of
contention. A coalition of environmentalists, lawyers, journalists and nongovernmental organizations has called for the release of the report as well
as public hearings on the project. They have cited a 2003 environmental
law that required public participation, including hearings, in deciding such
major projects.... The Chinese government has refused to release the report and, as yet, has not called any hearings. *Wen Wei Po* said the Ministry of
Water Resources and the State Secrets Bureau have classified the report as a
state secret, citing laws restricting the release of information about projects
on international rivers.[62]

On February 22, 2006, SEPA released two provisional measures to allow
for public participation in the environmental impact assessment process, stating that the public "may take part in the EIA by answering EIA
questionnaire [*sic*], consulting experts, or participating in a symposium or
public hearing." Equally important, developers would have to release the
data on the environmental impact of the project and the measures that are
in place to handle it. According to Pan Yue, "This is the first official document on public involvement in the environmental sector, which will make
government decisions in the sector more transparent and democratic." He
went further, asserting that the document would act as a guide for establishing a comprehensive system for releasing environmental information and
set into place mechanisms to ensure more public participation in the policy
process.[63] A month later, the first such request was formally registered at
SEPA.[64]

Some activists in Beijing remain pessimistic. One told me that a public
hearing on the NRP appears to be increasingly unlikely. Moreover, as of
March 2006, the report on the decision that was leaked to the *Wen Wei Po*
and discussed in the *International Herald Tribune* piece cited above has not
been disclosed publicly in China. In short, there is uncertainty and confusion
everywhere along the lines of debate. There is talk about the original thirteen dams being scaled down to only four (Maji, Liuku, Yabiluo, and Saige),
but there is always the chance that even if this is true, the four dams are
only "phase one" of a larger, as yet undisclosed project—the full implementation of the current NRP.[65] Indeed, the way the proposed four dams are

62. Ibid.
63. Sun Xiaohua, "Public to Help Assess Impact on Environment," *China Daily*, February 23, 2006.
64. Elaine Wu, "Activists Seek Hearing on Dam Project," *South China Morning Post*,
March 21, 2006.
65. Interview 06BJ01, March 10, 2006.

Figure 5.3 Surveying equipment, Songta section of the NRP, Songta, Tibet. Photograph by the author, March 2006.

laid out, they seem to be perfect candidates for a "foot in the door" type of project.[66]

Meanwhile, work continues apace. As part of my field research in March 2006, I visited various sites along the NRP, including the northernmost Songta Dam in Tibet, and preliminary surveying work was evident, even though this is not one of the four sites in the revised, pared-down proposal. This is confirmed by engineers in Maji, who said that such work is basic geological surveying work and that a final decision remains far away.[67] The following month, as a photo exhibit on the Internet attests, a visit by UNESCO officials to the Nu River demonstrated that some of these preliminary survey sites had been hastily "disguised" or had the more obvious elements (for instance, Guodian banners) removed.[68]

66. See Lampton, "A Plum for a Peach," in *Bureaucracy, Politics, and Decision Making in Post–Mao China,* ed. Lieberthal and Lampton, 33–58; and David Stanway, "Develop and Be Dammed: China to Build on Virgin River," *Interfax,* March 15, 2006. See also "China's Water Resources Minister to Revise the Nu River Planning," *Wen Wei Po,* March 6, 2006.

67. Interview KM/BJ01, March 18, 2006.

68. Available at www.threegorgesprobe.org/tgp/NuRiverGallery2/index.html, accessed July 27, 2007.

Analysis of the Nu River Case

Policy Entrepreneurs

Two of the principal policy entrepreneurs for the NRP opposition were Yu Xiaogang and Wang Yongchen, although there were many others. These two, when compared to a potential policy entrepreneur, He Daming, who remained largely mute throughout the entire process, provide the parameters of policy advocacy among the opposition to the NRP.

He Daming provides an interesting contrasting case because of his caution and self-imposed lack of visibility. He did not seek the spotlight; indeed, he avoided it. As He was being watched by the local authorities who were waiting for him to make a mistake, he remained largely above the fray. One could argue that he was indirectly supporting the government because he has amassed an unbelievable amount of scientific information on the Nu River (although to be fair, He has been prolific in publishing mostly technical articles and book chapters that are subtly critical of the planning process for hydropower projects throughout Yunnan), but he has largely sat on it. He has avoided writing the many papers that such information would enable, and he has shown a singular unwillingness to share the vast sums of data that he has collected over the past two decades. Yet he is perceived as a threat to the government for at least two reasons. First, he is a local and has a tremendous amount of legitimacy—as well as, reportedly, political backing at the national level. Second, by staying outside of the political struggle over the NRP, He can be seen as lacking an agenda; he simply has the empirical data upon which he makes his own conclusions, based on his scientific assessment. Such credibility is what threatens the proponents of the NRP and is precisely what Fang Zhouzi attempted to undermine in his lecture at Yunnan University in April 2005.

Among the actual policy entrepreneurs, Wang Yongchen has been pivotal to the twists and turns of the Nu River controversy. She entered the fray not as an expert but as an environmental activist from faraway Beijing. Nevertheless, she established her bona fides by making multiple trips to the Nu River Valley, often with a group of journalists. She also kept the issue alive through the monthly salons she held:

Wang Yongchen's Green Earth Volunteers journalist salon holds a meeting every month. Wang says that because of this, real environmental protection activists can use the media and these environmental protection organizations have started to have some influence. Generally speaking, everybody in the salon has two identities: one is as a journalist or as a government official, and the other is as an environmental protection volunteer.[69]

By also dipping into her own personal savings to publicize the plight of the *yimin* and the beauty of the Nu River Valley through the March 2004 photo exhibition in Beijing, Wang was further able to influence public opinion where it mattered, as far as expanding the sphere is concerned: not at the local level but at the national level. Wang's role in providing information as well as framing the contours of the story for the multitude of journalists of her acquaintance cannot be overstated.

Yu Xiaogang is perhaps the most fascinating of the policy entrepreneurs, in part because he appears to be so willing to take on risk. He has engaged in activities that consistently embarrass the authorities, and although he is now closely watched and unable to travel abroad, he continues to travel domestically and to press the issue. What accounts for Yu's propensity to take such daring political risks? Part of it has to do with individual idiosyncrasies, but that does not get us very far. Equally important is that, as with others discussed in previous chapters, it is difficult to touch Yu politically. This has less to do with any formal position or *guanxi* (informal reciprocal power relationships), although that is certainly a factor (he is married to the daughter of a former general). Rather, it has to do with Yu's quite genuine revolutionary credentials, something that sets Yu apart from others in his age group.

Like many of his generation, Yu was sent down to the countryside during the Cultural Revolution, in his case to Yunnan. However, once he found himself working with the peasants, he broadened his horizons by crossing the border to Myanmar, fighting alongside the Burmese communists against the forces of Ne Win. By any objective standard, Yu has revolutionary credentials that far outweigh his generational peers. One might even say that Yu exhibits exactly the type of political background that provided the CCP with its tremendous legitimacy when it took the reins of power in 1949. He is not untouchable, but actually "touching" him requires a tremendous amount of planning and the existence of a "smoking gun" (generally some sort of illegal activity), which Yu is extremely careful to avoid. Yu's activities focused on providing information to those who questioned the hydropower projects in Yunnan, and although Yu's conclusions were often quite measured, his methods have been unprecedented in contemporary Chinese political discourse.

69. Pubugou Interviewee, July 8, 2005.

Issue Framing

Three factors ensured that there would be extensive media coverage on the Nu River controversy: the constellation of NGOs that were active in opposing the NRP, many NGO members who are former or current professionals within the media (journalists and editors), and the clear "us versus them" aspect of the story. That media coverage was critical; in its absence, it is likely that the NRP would have gone ahead as planned with relatively muted opposition, if any. The Nu is in a very remote part of China; because of its relative isolation, local officials who were very much in favor of the project were able to maintain a relatively tight monopoly of information. The coverage of the Nu went beyond simply proposing or opposing the NRP but often took the form of laying out several competing opinions. Although many of the issues that were at stake found their way into the story, the main focus became that of environmental protection.[70]

Between August 2003 and September 2004, there were more than one hundred domestic and international media stories on the Nu River as well as hundreds of reports and surveys. There were four important dimensions to the way in which the media covered the Nu River controversy. First, breaking from tradition, the media evolved from simply supporting such projects to documenting the opinions on all sides of the issue. Second, the media provided more analysis of the situation, eschewing the superficial approach that had been the standard of reporting. Third, the media was able to move public opinion away from the "economic development is good for the western part of China" party line and toward such notions as respecting cultural heritage (Yangliuhu/Dujiangyan) and biodiversity (Nu River). Finally, and most important, the media provided information. Insofar as the biggest obstacle for public participation in the political and policy process is the scarcity of information, the media sought to fill this gap.

Of course, given the issues at stake for all sides of the NRP, it is difficult to argue, as some have, that all sides of the issue were generally objective. In fact, while much of the information was straightforward journalism, some

---

70. See, among others, "*Nujiang daba gongcheng zanhuan beihou de minjian liliang*" [The People Power behind the Stoppage of the Nu River Project], *Weekly News* (*Xinwen zhoukan*), May 20, 2004, available at news.sina.com.cn/c/2004-05-20/15043285303.shtml; "*Nujiang daba "chusheng" you dian bing*" [The Somewhat Sickly Birth of the Nu River Project], *Morning Report* (*Xinwen chenbao*), November 26, 2003, available at news.sina.com.cn/c/2003-11-26/03001187537s.shtml; "*Zhongguo NGO: wo fandui*" [Chinese NGOs: I Oppose!], ChinaNewsweek.com, July 9, 2004, available at www.chinanewsweek.com.cn/2004-07-09/1/3855.html; and "*Renmin huanbao yishi suxing Nujiang daba jihua zaofandui*" [Reemergence of Popular Environmental Consciousness, Opposition to the Nu River Project], *Epoch Times* (*Dajiyuan*), December 30, 2005, available at www.epochtimes.com/gb/5/12/30/n1172454.htm, all accessed May 29, 2007.

of the coverage took on the trappings of official media (supporters) and of opinion columns or even tabloids (opponents). But perhaps this is looking at the issue the wrong way; what is significant is that the media were able to shift, qualitatively if not quantitatively, the asymmetry of the coverage of the NRP from economic development to environmental and other social issues in a way that brought widespread societal interest to an otherwise esoteric and inaccessible (to most Chinese, anyway) story.[71]

The frame that caught the attention of the public was that of the environment. Others have argued that in spite of China's continuing development, environmentally friendly "postmaterialist" tendencies have arisen. This is partly due to official propaganda, part of it is simple aesthetics, and part of it arises from a traditional Chinese view that humanity is very much subordinate to nature, a theme echoed in countless Chinese paintings and poems over the past two millennia.

But if the frame of economic development was insufficient to keep the alternative frame of environmental concerns from altering the contours of debate, the pro-NRP forces did have an effective counter-strategy: scientific assessments. This was able to undermine the credibility of the opposition in two ways. First, the Achilles' heel of the environmentalists has always been their emotive stance on the state of the environment. Although often an extremely effective tool in mobilizing public interest, it is easily attacked by those who can demonstrate that *their* view, based on "objective, scientific" criteria, is more in line with the facts of the case. This resonates with the Marxist undertow of Chinese political discourse, since Marxism has always been regarded as a "scientific" theory. Moreover, the actual language used by some of the environmentalists, such as by Wang Yongchen in her response to He Zuoma, for example, has played right into the dam proponents' hands.

The second way in which this alternate objective scientific frame has hurt the opposition is in the information on which it is based, which is often *neibu*, or classified or "for internal use only," making it impossible for the anti-NRP forces to provide counter-arguments based on the "scientific merit" of this case. This is precisely why there has been such unrelenting pressure on He Daming, since He actually could provide such data. In a sense, this strategy is exactly the opposite of "expanding the sphere"—while the NRP opponents try to increase the information that supports their point

---

71. *"Nujiang jianba zhi zheng yu zhuanmo xinzi de duichen"* [The Symmetry between the Nu River Controversy and the Media Reportage of It], *Scientific Valley Development* (Kexue fazhan guan yu jianghe kaifa), Beijing, China: Huaxing chubanshe, 2005, 141–52.

of view, the NRP proponents jealously guard it. Indeed, they not only keep such information from the opposition, they also keep such information from real and potential allies, such as local officials whose jurisdictions line the banks of the Nu River itself. It may be a cliché, but knowledge really is power, and the NRP proponents have maintained this relative monopoly of information, closed ranks, and frustrated the opposition's desire to debate the issue on these terms.

For example, in late October 2005, the China Rivers Network, a consortium of seven NGOs, boycotted a government-organized conference. Their objections were over the fact that the environmental impact assessment report would not be made available to the public (and, by implication, leaking the contents might be construed as passing on "state secrets") even though they had been assured that they would have access to the report. Zheng Yisheng, a researcher at the Center for Environmental and Development within the Chinese Academy of Social Sciences, complained: "The organizers said they would share with us parts of the environmental impact assessment. *But we don't want private access to the documents.* Why not make them accessible to everyone?"[72] Insofar as the NRP opponents are unable to gain traction on the scientific front, they are even more overly dependent on getting the message out by framing the issue not only in environmental terms but also in environmental terms that they hope will link these issues directly to the preferences of their target audience, whether emotive or not. In a sense, for the NRP opponents, given their weakness on the scientific front, their strategy has been to overwhelm the opposition by attempting to enlarge and to expand the sphere as much as is politically possible.

Finally, it is often very difficult to disaggregate the complexity of issues involved and advocate an environment-friendly position that does not somehow underreport the significant opportunity costs involved. For instance, hydropower is an extremely clean form of renewable energy that mitigates China's continuing reliance on the decidedly environmentally deleterious resource of coal. Indeed, as I will argue in the concluding chapter, an unintended consequence of this increased political pluralization of environment-related issues may well be to constrain China further in its choice of energy resources for future development.

Broad Support for Policy Change

Nevertheless, the environmental frame remains a strong one. In China, citizens and politicians alike have been bemoaning the negative environ-

72. Bezlova, "Let Public See Secret Mega-Dam Plans" (my emphasis).

mental consequences brought about by the rapid transformation of the economy under reform, which followed three decades of Soviet-style environmental degradation unleashed under Mao.[73] One of the strengths of this frame is that it is ostensibly nonpolitical. It is a goal compatible with any regime type, at least in theory. Rather, it stands in opposition to particular industrial and economic strategies. Moreover, the environmental frame is not only compatible with numerous laws and regulations already on the books, it consciously calls for the enforcement of these very laws and regulations. As a result of such political cover, it is a frame uniquely poised to mobilize the citizenry on a moral high ground that will prove very difficult for the authorities to ban outright.

Thus, the potential for expanding the sphere is substantial. In addition to the Chinese citizenry and officials for whom environmental concerns mesh seamlessly with their political goals, there is a wider audience as well. Although it is important not to overstate the direct impact of international opinion, it is equally important not to underestimate its indirect and symbolic impact and the very concrete results that flow from it. First of all, as was demonstrated in the Dujiangyan case in chapter 4, Beijing is quite sensitive at being labeled a regime that is unfriendly to internationally accepted norms, such as World Heritage protection. Although UNESCO is a weak player in the hardscrabble world of Chinese politics, its mantle is an effective card for the opposition to invoke in its campaign against the "develop at all costs" strategy of the Yangliuhu proponents. This is also true with regard to environmental issues.[74]

Second, insofar as pollution gets at the very notion of negative externalities, Japanese and Korean protests over the acid rain emanating from Chinese industrialization dovetails perfectly with the anger that many Chinese feel when their downstream communities are contaminated by upstream industries that reap the benefits of industrialization without sharing the costs.[75] That is to say, the international dimensions of China's environmental problems resonate with some of the very same issues that face the Chinese state domestically. Thus, China's leaders feel boxed in from two sides on this issue, and activists can use the international aspect of China's

---

73. Judith Shapiro, *Mao's War against Nature: Politics and the Environment in Revolutionary China*, New York: Cambridge University Press, 2001.

74. See, for example, UNESCO, "State of Conservation (Three Parallel Rivers of Yunnan Protected Areas)," Decision 30 COM7 B. 11. See whc.unesco.org/en/decisions/1094, accessed July 29, 2007.

75. David Lague, "China Blames Oil Firm for Chemical Spill," *International Herald Tribune*, November 25, 2005.

environmental situation to underscore its problems at home with less risk of sanctions from the authorities.

The very fact that activists like Ma Jun and Yu Xiaogang have received accolades from abroad is another potential embarrassment to Beijing—in 2006 Ma was named one of *Time* magazine's 100 most influential people, and Yu won the prestigious Goldman Prize.[76] Not only does it bring unwanted attention to a state that is not doing enough to stem the tide of environmental degradation—that is, it falls upon the relatively powerless "Davids" to take action—but it also provides a degree of political cover for these individuals so that they cannot be "removed" from the limelight they currently enjoy without placing Beijing in a considerably embarrassing position.

The Nu case is also important conceptually because it is likely the most representative of the new politics of hydropower analyzed in this book. While Dujiangyan was an unqualified success in the attempts to pluralize the political process and Pubugou was an unmitigated failure in the same regard, the Nu case contained far more loose threads, debates, and unresolved conflicts. The Nu case represents a more multifaceted and multilayered set of political and policy dynamics with regard to expanding the sphere of conflict. The political and policy implications of these three cases on China's domestic and international profiles form the basis of the final, concluding chapter.

76. Ed Norton, "Ma Jun," *Time,* April 30, 2006, available at www.time.com/time/magazine/article/0,9171,1187271,00.html; and Goldman Environmental Prize, "Yu Xiaogang," available at www.goldmanprize.org/node/443, both accessed May 30, 2007.

6 | A Kinder, Gentler "Fragmented Authoritarianism"?

> These farmers are angry that plans are being made to dam the Yangtze River,
> flood Tiger Leaping Gorge and force the relocation of thousands of farmers
> and villagers. And they are getting vocal, learning about their legal options
> and pressing local officials to reconsider how the dam will be built. Getting
> political is not a hobby for these farmers. It is a necessity.
>
> And similar dramas of necessity are being played out all over the Chinese
> countryside today by villagers who know that they are not fully participating
> in China's economic growth, but are being told that if they want to, they must
> accept dams or factories that will destroy their environment.
>
> They don't like this deal, but China's rigid political system leaves these
> farmers, who are still the majority in China today, with few legal options for
> fighting it. That helps explain why China's official media reported that in
> 1993 some 10,000 incidents of social unrest took place in China. Last year
> there were 74,000.
>
> This is the political lens to watch China through today. How China's ruling
> Communist Party manages the environmental, social, economic and
> political tensions converging on such places as Tiger Leaping Gorge—not
> Tiananmen Square—will be the most important story determining China's
> near-term political stability.
>
> —THOMAS FRIEDMAN, "How to Look at China," *New York Times*,
> November 9, 2005

Columnist Thomas Friedman lays out a somewhat dramatic scenario
that, while accurate in sketching out the sources of political tension, is
incomplete and quite possibly wrong in its implied conclusions: the political implications of the cases analyzed in this book are not explained by
variables conceptually associated with protests, democratization, or continued static authoritarian rule. Rather, they infer subtle shades of political
pluralization in China. This book suggests that the dichotomy surrounding the debate over political liberalization—top-down versus bottom-up
democratization—might not be the most effective metric with which to
measure a country's progress in eliminating traditional authoritarianism.
Rather, what we are seeing in Chinese hydropower politics is a type of

pluralization in which very real and substantive participation by actors hitherto forbidden to enter the policymaking process—NGOs, the media, and disgruntled opponents inside and outside of the government—are now increasingly able to do so. This is because they do not threaten—and are not seen as threatening—the legitimacy of the government or the Chinese Communist Party. Indeed, many of these activists are themselves CCP members. What they do threaten is the inevitability of policies that activists deem to be detrimental to various aspects of Chinese society—and to themselves.

Democracy vs. Political Pluralization

Recent literature on democratization in China tends to focus on one of two dynamic processes: elite-driven change and grassroots efforts. Gazing into his crystal ball, Bruce Gilley ventures that democratization is likely to be elite-initiated in the form of a rational response to a "multiple metastatic dysfunction." Apart from the unfalsifiability of such claims—at least at present; we have to wait several decades—Gilley assumes that elite-level democratization at the national level would necessarily lead to replication of such events at the local level throughout China. In fact, much of the top-down literature on democratization suffers from the same logical dead-end: that democratization imposed on local governments from on high in Beijing is not really democracy at all.[1]

Much of the literature on the mechanisms of democracy in China, particularly those works focusing on township and village elections, is generally of a very high caliber, yet it remains somewhat tentative and subject to a number of methodological and political constraints. As a result, we are left with two bookends, both of which may be missing an important dynamic: the pluralization of the Chinese political process.[2]

It is my belief that focusing on democratization in China has prevented us from understanding the degree of political liberalization that is taking place right in front of our eyes. But if we shift our conceptual lens away from democratization and analyze instead political pluralization—that is, the increased role in the policy process by individuals and groups both inside

1. Bruce Gilley, *China's Democratic Future: How It Will Happen and Where It Will Lead*, New York: Columbia University Press, 2004.

2. Melanie F. Manion, "The Electoral Connection in the Chinese Countryside," *American Political Science Review* 90, no. 4 (1996): 736–48; and M. Kent Jennings, "Political Pluralization in the Chinese Countryside," *American Political Science Review* 91, no. 2 (1997): 361–72.

and outside the traditional arenas of policymaking—the picture changes significantly.

Of course, I am not suggesting that that these changes have entirely supplanted the dominant paradigm of zero tolerance for dissent. Important limits on the state's willingness or capacity to accept the demands of hydropower and dam opponents combined with the corrosive effect of local corruption on the political and social capital necessary to reach compromise can lead to a stalemate, held together by state coercion with the specter of widespread social unrest lurking just around the corner. The prospect of such unrest keeps China's leaders awake at night and pushes their policy and political preferences further in the direction of maintaining social stability above all else—to which the case of Pubugou is a grim and sobering testament.

Protest in China

Casual observers of China were shocked when they read about police officers firing on protesters in Dongzhou township in Guangdong province in early December 2005. No doubt this was because the reports suggested that some people had been killed. But it was also due to the locale: Guangdong, just across the border with Hong Kong and a symbol of China's economic liberalization and political pragmatism.[3] For those who had been following China more closely, however, these protests illustrate a dramatic trend: the mushrooming of demonstrations in China.

The State Security Bureau has released figures, now widely cited, of 58,000 protests in 2003, 74,000 in 2004, and 87,000 in 2005.[4] There has been some scholarly debate over the significance of these numbers. Some argue that these figures are relatively low when you look at China's overall population or that only a few of these are large-scale organized protests as we might imagine them in the West, leaving the majority of them quite small and therefore insignificant.[5] Others argue that protesters are hindered by the fragmentation of the workforce, the poor prospects for workers disenfranchised by

3. Edward Cody, "Chinese Police Kill Villagers during Two-Day Land Protest," *Washington Post*, December 9, 2005; and Howard W. French, "China Calls Clash Result of 'Chaotic' Mob Attack," *New York Times*, December 11, 2005.

4. Joseph Kahn, "Pace and Scope of Protest in China Accelerated in '05," *New York Times*, January 20, 2006.

5. Pierre F. Landry and Tong Yanqi, "Disputing the Authoritarian State in China," paper presented at the annual meeting of the American Political Science Association, Washington, D.C., September 1, 2005.

Table 6.1 Incidents of social unrest in China, 1993–2005

Year	Number of protests	Percent change
1993	8,700	N/A
1994	10,000	14.9
1995	11,500	15.0
1996	12,500	8.7
1997	15,000	20.0
1998	24,500	63.3
1999	32,500	32.7
2000	40,000	23.1
2001	N/A	N/A
2002	50,400	N/A
2003	58,000	15.1
2004	74,000	27.6
2005	87,000	17.6

Sources: Albert Keidel, "The Economic Basis for Social Unrest in China," Third European-American Dialogue on China, George Washington University, Washington, D.C., May 26–27, 2005; and Joseph Kahn, "Pace and Scope of Protest in China Accelerated in '05," *New York Times*, January 20, 2006.

reform, the strong deterrent measures for potential protest leaders, and the strategic dismantling of the state-owned enterprise sector to minimize social instability (or, rather, to isolate into smaller groups those laid-off workers who would be most likely to lead a protest).[6]

Another group of scholars has focused on the process of the protests themselves and how protesters can come away with something, even if they lose. In their work on "rightful resistance," O'Brien and Li argue as follows:

> Rightful resistance is a form of popular contention that operates near the boundary of authorized channels, employs the rhetoric and commitments of the powerful to curb the exercise of power, hinges on locating and exploiting divisions within the state, and relies on mobilizing support from the wider public. In particular, rightful resistance entails the innovative use of

6. See, among others, Cai Yongshun, *State and Laid-Off Workers in Reform China: The Silence and Collective Action of the Retrenched,* London: Routledge, 2005; Ching Kwan Lee, "From the Specter of Mao to the Spirit of the Law: Labor Insurgency in China," *Theory and Society 31,* no. 2 (April 2002): 189–228; Hurst and O'Brien, "China's Contentious Pensioners"; and William Hurst, "Understanding Contentious Collective Action by Chinese Laid-Off Workers: The Importance of Regional Political Economy," *Studies in Comparative International Development* 39, no. 2 (Summer 2004): 94–120.

laws, policies, and other officially promoted values to defy disloyal political and economic elites.[7]

The phenomena I analyze share some elements with this concept of "rightful resistance," but they differ in some important ways. The differences mainly hinge on the notion of resistance. In the cases that I document, opponents to state policy are not simply content with resisting policies that affect them directly; they also seek to change the substance of broader policies. Second, a related point, their principal targets are not merely those local officials whose corruption and other malfeasance run counter to legal and other norms. Rather, their focus is on entering in and working within the policy process to meet their principal policy-related goals. In fact, the one case in this book that is closest to O'Brien and Li's framework, Pubugou, is notable for its failure in the policy realm.

On the other hand, other recent work on protests in China suggests an analogue with the notion of official state institutions or actors that are nonetheless critical allies of the opposition. As Perry and Selden write, "While local officials frequently crack down on popular resistance, in numerous cases their leadership is instrumental in shaping, legitimating and articulating the demands of social movements, and in some cases in networking with state officials on behalf of local interests."[8]

But "protests" per se are not the focus of this book, because as the Pubugou case illustrates, the powers that be can always point to widespread protest in the policy issue at hand and invoke the importance of maintaining social stability and summon the specter of state coercion. As a result, hydropower opponents are as leery of widespread protest as the authorities are: they know that such protest will undermine their policy and even their normative goals. Many of these actors cut their teeth during the 1989 protests at Tiananmen and elsewhere, and they experienced firsthand the power of the state in quashing organized opposition to state policy. Rather like their counterparts in the United States and Europe in the years following the 1960s, some of them seem to have decided that if they want to change state policies, they have to become part of the system. What is remarkable is that they have been able to do just that within the fissures of China's fragmented authoritarian political apparatus.

7. Kevin J. O'Brien and Lianjiang Li, *Rightful Resistance in Rural China,* New York: Cambridge University Press, 2006, 2.

8. Elizabeth J. Perry and Mark Selden, "Introduction: Reform and Resistance in Contemporary China," in *Chinese Society: Change, Conflict, and Resistance,* ed. Perry and Selden, 2nd ed., New York: RoutledgeCurzon, 2003, 11.

Implications for Domestic Politics in China

In their seminal work on policymaking in the 1980s, Lieberthal and Oksenberg describe the dynamics of the policy process in China, a framework they describe as "fragmented authoritarianism":

> Policy X resulted from a bargain among Ministries A, B, and C and Province D either 1. brokered by one or more top leaders, 2. arranged by coordinating staffs acting in the name of one or more top leaders, or 3. negotiated by the supra-ministry coordinating agency, and ratified through routine procedures by the top leaders. Disgruntled Ministries E and F, losers in the deal, planned to pursue strategies to erode the agreement. The bargain sought to reconcile the conflicting organizational missions, ethos, structure, and resource allocations of the ministries involved.[9]

Nothing in this quotation is inconsistent with the analysis contained herein, but it remains incomplete. It does not capture some of the key dynamics that have been presented in this analysis. While at first glance, one might argue that the opponents of these hydropower projects are simply acting out the traditional roles of disgruntled "veto players" in the political process, this is inaccurate. The opposition, even when its members are contained within various local bureaucracies, often does not have the veto power that implementation bureaucracies in the fragmented authoritarianism (FA) framework possess. On the other hand, its goal in the policy process is far less modest than that of a mere spoiler: the goal is not to undermine the policy through neglect but to reverse the policy to contribute to a recasting of it that takes their considerations into account.

But this does not tell us much about how these policy opponents, bereft of such veto power, are able to amass the requisite power to pursue their goals in a political system in which very often "might makes right"—at least in the form of formal and informal power. In this instance, the FA model is instructive. First, fragmentation provides fissures in which one of the most important aspects of power—information—is jealously guarded. Such an environment decreases the amount of information available, increasing its value. Thus, previously unavailable information has a much greater impact than it might in a situation in which information was freely available all along. The leaking of information by local agencies to journalists, combined with journalists' own instincts (and increasingly, mandate) for information-gathering, takes on a significant degree of political importance.

---

9. Kenneth Lieberthal and Michel Oksenberg, *Policy Making in China: Leaders, Structures, and Processes,* Princeton, NJ: Princeton University Press, 1988, 4.

Second, as Lieberthal asserts, "The encouragement given to many organs to become increasingly self-supporting through bureaucratic entrepreneurship," a dynamic that I have described in the intellectual property policy arena, has also "strengthened the tendency of bureaucratic units to work vigorously to promote and protect their own interests in the policy-making process."[10] The legitimacy given to bureaucratic and organizational activism has animated those units at the very bottom or along the sidelines of the system and made them increasingly important players in the process.

Third—a related point—within a fragmented political system, argues Lieberthal, "policy communities" are an important source of information and strategic coordination of discrete and sometimes disparate actors into a coherent group. However, this is not at all limited to the status quo powers; in recent years, this has provided power and leverage to the hitherto powerless opposition.

Indeed, Lieberthal expounds on one of these dimensions not captured as explicitly in *Policy Making in China*, that of value integration. He describes the importance of ideology during the Mao era in maintaining a shared perspective of officials in diverse geographical locations (sometimes with only sporadic communication with one another). Although the analysis contained herein is consistent with Lieberthal's claim that "shared values...can substantially affect the operations of a political system," it suggests that the shared values of the hydropower opponents might be stronger (or at least more prone to distributional efficiencies among potential allies and new recruits) than that of the dam proponents. This is captured in the linchpin of expansion of the political sphere of conflict in China: issue framing. Rhetoric aside, the benefits of hydropower appear to be much more of a private good in practice, while preservation of an alternate, oppositional frame is more of a public good in which policy entrepreneurs are able to overcome collective action problems by taking on responsibilities that the vast majority of their supporters may not want, or are in no position to undertake in any case. Again, to quote Lieberthal: "Value consensus can basically reduce the need of the political leadership to develop additional resources to assure fidelity to their priorities and compliance with their policies."[11] This works both ways, that is, it provides the policy entrepreneurs with the political leverage necessary to successfully pursue their goals, as

10. Kenneth G. Lieberthal, "Introduction: The 'Fragmented Authoritarianism' Model and Its Limitations," in *Bureaucracy, Politics, and Decision Making in Post-Mao China*, ed. Lieberthal and Lampton, 9.

11. Ibid., 7.

the successful campaign at Dujiangyan and the protracted debate over the Nu River Project demonstrate.

In describing the policy dimension of the FA framework, Lieberthal and Oksenberg conclude that

> policies are not necessarily either coherent or integrated responses to perceived problems or part of a logical strategy for a leader to advance power and principle. For analytical purposes, a package or bundle of policies is best disaggregated into its individual or separate policies. While some of those policies may result from the initiative of top leaders, others are best seen as a temporary agreement arranged by the top leaders among contending and powerful bureaucracies with diverse purposes, experiences, and resources.[12]

It is important to note that none of these cases provides an instance in which a component of a larger policy, the campaign to "develop the West" (*xibu da kaifa*), was taken on directly by hydropower opponents. To do so would almost certainly have resulted in failure. Moreover, and far more important, the opponents to particular projects are not necessarily against the overall policy goals of developing the West. Rather, they take issue with certain component parts of the policy package. By doing so, they disaggregate the overall policy "line" into discrete parts that are manageable enough so that there is a decent chance of reshaping the policy in ways that are consistent with their goals.

Thus, rather than indicate the obsolescence of the FA framework, the research contained in this book suggests that nontraditional members of the policymaking process in China, local (that is, subprovincial) officials, the media, NGOs, and individual activists have successfully entered the political process precisely by adopting the strategies necessary to work within the constraints of the FA framework.

Of course, one can argue that this is simply a cooptation of the opposition by the state and that our optimism about political liberalization in China must be tempered accordingly. In a sense, this could be described as akin to a more inclusive form of neo-corporatism.[13] Yet what I am describing exists along a continuum: the increasing levels of inclusion at some point reach a tipping point in which an ever-growing corporatist system begins to resemble a broader heterogeneous system of competing interests for parts

12. Lieberthal and Oksenberg, *Policy Making in China*, 4.
13. See, for example, Philippe C. Schmitter, "Still the Century of Corporatism?" *Review of Politics* 36, no. 1 (January 1974): 85–131; and Kenneth W. Foster, "Associations in the Embrace of an Authoritarian State: State Dominion of Society?" *Studies in Comparative International Development* 35, no. 4 (Winter 2001): 84–109.

of the policy pie. Although caution is warranted, it appears that China, as a dynamic and evolving political system, is not likely to remain wedded to any single state ideal type, so that even if events I have described in this book might be interpreted today as a form of corporatism, it seems unlikely that they will remain so into the future.

Moreover, as the contributors to the volume *Grassroots Political Reform in Contemporary China* suggest, grassroots initiatives, regardless of whether they are initiated by the local state or Party apparatus or by society, have nevertheless contributed to a more dynamic and diverse political process. Thus this somewhat pessimistic invocation of cooptation, I believe, sets up a false dichotomy in terms of the evolutionary processes between state-initiated and more traditional society-initiated grassroots efforts to influence policy outcomes and ignores the evolutionary processes I am suggesting are taking place.[14]

In short, I agree that we should be modest in what we expect as far as political liberalization is concerned. However, an absolutely critical component of politics—the barriers to entry into the political process—arguably separates liberal political regimes from illiberal ones, and they have been demonstrably lowered in China with regard to hydropower policy. One may object to my conceptualization of policy entrepreneurs as overly strategic, but all that the characterization really means is that these individuals have learned the lessons of how to prevail—or at least compete—in the hardscrabble world of Chinese politics without having lost their normative and often personally meaningful policy goals.

In a sense, this means that the pluralization of the policy process in China provides both less and more influence on policy than would certain forms of democratization. On the one hand, meaningful elections, the mechanism on which many definitions of democracy depend, do not exist in China, except at village and township levels (and these are widely understood to be designed to enhance CCP control through the election of the most "popular" Party officials on the ballot). On the other hand, the role of the opponents to these hydropower debates and to the actual politics of hydropower is both direct and firmly embedded within the political process—as much as is the case in many democratic regimes, if not more so. Of course, one must be careful in taking this implication too far; rather, the more appropriate claim here is that while the FA framework continues to define the major contours of the policymaking process in China, the

---

14. Elizabeth J. Perry and Merle Goldman, eds., *Grassroots Political Reform in Contemporary China*, Cambridge, MA: Harvard University Press, 2007.

number and types of actors within it have increased dramatically in recent years.

It would be remiss of me not to acknowledge that some have argued that the FA framework is flawed and could even point to the examples cited in this book for support. For example, Hamrin and Zhao argue that the very articulation of "fragmented authoritarianism"

> leaves something to be desired in that it implies an uncontrolled process whereby autonomy is being seized by subunit actors against the will of the state. This does not evoke the state's complicity in its own devolution and overstates the weakness of the center. The focus on economic dynamics leads to a rather vague conception of the state structure, as reflected in the suggestion that China is in transition from a traditional hierarchical system toward a more modern, market-oriented system.[15]

There are several problems with this argument. First, there is nothing within FA or the analysis contained herein that suggests that the state is complicit in its own devolution. Rather than complicity, what seems to be occurring is that the state increasingly recognizes its own limitations in objective terms. Rather than simply pointing at the "weakness" of the center, the cases in this book point to significant ambivalence with regard to the multidimensional aspects of hydropower policy.

Second, Hamrin and Zhao misinterpret the notion that FA is premised on economic issues in their assertion that it leads to a "vague conception" of the structure of the state. The problem with the FA framework is not vagueness but rather that the FA framework itself is static. It was not a dynamic, evolving framework when it was first articulated in the 1980s and early 1990s. However, by superimposing this study on the earlier work by Lieberthal and Oksenberg, and others, we can add a longitudinal dimension to see how the FA framework has evolved since then.

Implications for China in the International Arena

This book has focused almost exclusively on domestic processes. However, as a student of international relations, I see no reason to limit the implications of this analysis to domestic politics. In contrast to previous analyses of second-image reversed framework—that is, how the international system affects domestic politics—in this study it appears that the second image proper—how domestic politics influences a state's international behavior—is not only appropriate but vital to understanding China's international

15. Carol Lee Hamrin and Zhao Suisheng, eds., *Decision-Making in Deng's China,* Armonk, NY: M. E. Sharpe, 1995, xxvii.

behavior, particularly with regard to the growing concern over the international politics of energy.

Although there is much debate over the intersection between domestic politics and international relations, there is a tendency to treat these two arenas as separate domains of scholarly inquiry. When I began this research, I fully expected to draw from both of these subfields of political science in fashioning an explanation for the empirical puzzle that drove this research: the abrupt shift of policy as a result of widespread opposition. Although international variables have played a role, particularly in providing valuable information to the participants and in providing forums for the articulation of their goals (mainly through the channels of international organizations), I was surprised to find that the majority of the political processes described in this book are largely contained within China.

This does not, however, mean that the political and policy implications of these processes are similarly restricted to domestic politics in China. Indeed, by looking through the conceptual lens of the second image, the international implications of these domestic processes are nothing short of profound. These implications revolve around the fact that energy resources are scarce and that the constraints on exploiting them are high. Students of democratization often lump together the notion that green movements and democratization are mutually reinforcing elements that lead to more environmentally friendly policies of the states in which they operate. But this ignores the role of opportunity costs.

China is in a somewhat interesting position in that it possesses ample coal reserves.[16] That means that politically it may be constrained in terms of a large portion of its energy sources but also that it could always maintain its fallback position of increasing its reliance on coal (putting aside for a moment the negative economic impact of pollution). The oppositional issue frame in the Nu River case, environmental degradation, would seem to seek to limit China's coal consumption as well. And the hydropower opponents are not against economic development per se but rather against the negative impact of economic development on China's other resources. But this unstable package of policy preferences—otherwise referred to as "social traps"—is internally contradictory. That is, insofar as China limits its exploitation of hydropower, it increases its tendency toward reliance on coal and other nonrenewable sources of energy. The irony here is that the political pluralization of China's hydropower policy may well impose more—not fewer—constraints on China's energy choices. And this would lead to

16. See www.cslforum.org/china.htm, accessed July 27, 2007.

a perverse outcome: an increased reliance on precisely those sources of energy that are anathema to environmentalists.

But this also bleeds into China's foreign policy, specifically, the international politics over energy. Policymakers in Washington anticipating increased conflict with China may well be looking in the wrong places. They may focus on the increases in China's defense spending as an indicator that China is developing a more aggressive military stance, even as they ignore that these budget increases are largely a result of the extensive divestment by the military of commercial enterprises, as mandated by Beijing.[17] Perhaps the areas where these same policymakers might draw some comfort are contained in the preliminary conclusions of this book—to wit, that China is becoming more pluralized in its policymaking process. They may even interpret this as a proxy for democratization. However, this is precisely the dimension where their overall trepidation over Sino-U.S. relations in the twenty-first century may be most justified.

Increased pluralization within the policy of hydropower in China leads to a narrowing down of energy choices in China. Because of the importance of energy to China's continued development, it would seem fair to assume that Beijing will curtail the role of these oppositional elements to hydropower policy. But as the cases of Dujiangyan and the Nu show, Beijing lacks the ability, the will, or both, to do so. As a result, there appears to be a growing number of hardening constraints on Beijing's energy choices, at least in the short to medium term. If we now bring in the negative economic effects of pollution on economic development, it becomes clear that the degree of China's reliance on coal will be difficult to sustain. If Beijing's constraints over its energy choices increase at home, it must either suspend its economic development (a nonstarter) or it must compete with corresponding vigor for energy sources abroad.

Thus, those who advocate increased democratization in China and who may welcome the types of political pluralization described in this book may one day wake up to witness a China that is more confrontational and less likely to be conciliatory, because China has been prevented from exploiting its own domestic sources of energy. This has potentially enormous implications for those who subscribe to theories of international relations that are driven by the assumptions of democratic peace.[18] That is, as China becomes

---

17. James Mulvenon, *Soldiers of Fortune: The Rise and Fall of the Chinese Military-Business Complex, 1978–1998,* Armonk, NY: M. E. Sharpe, 2001.

18. See, for example, Lisa L. Martin, *Democratic Commitments: Legislatures and International Cooperation,* Princeton, NJ: Princeton University Press, 2000.

less authoritarian, it may actually become more confrontational with the United States and other democracies over satisfying its growing energy needs.

China is already raising eyebrows in Washington because of its willingness to engage in friendly relations with the governments of Sudan, Iran, and other regimes with whom the United States refuses to deal, as a policy driven in part by China's energy needs. But, rhetoric aside, this is actually less likely to lead to confrontation insofar as the United States remains disengaged from these "rogue regimes." That is because such international engagement over energy is not necessarily a win-lose situation. The real difficulties will arise as China increasingly contends with the United States and other countries for energy from the latter's traditional suppliers of oil. With every day that passes, the international competition over sources of energy becomes more zero-sum.

Over the next ten years, it will be interesting to see how the Chinese political and policy processes evolve. The evolution of Chinese politics is not linear but somewhat dialectical (to use Maoist terminology). There may be situations in the near future where the political pluralization described in this book may be curtailed. However, such a development is likely to be temporary. The twin genies of policy entrepreneurship and issue framing have been let out of the bottle and have been allowed to seep into the fragmented policy making system. In subtle yet dramatic ways, Chinese politics are changing as a result.

Index

Page numbers in italics refer to figures and tables.

DATE DUE

PRINTED IN U.S.A.